Sustainable Construction Pr

Sustainable Construction Processes
A Resource Text

Steve Goodhew

**School of Architecture, Design and Environment
Plymouth University, UK**

WILEY Blackwell

This edition first published 2016
© 2016 by John Wiley & Sons, Ltd

Registered Office
John Wiley & Sons, Ltd, The Atrium, Southern Gate, Chichester, West Sussex, PO19 8SQ,
United Kingdom.

Editorial Offices
9600 Garsington Road, Oxford, OX4 2DQ, United Kingdom.
The Atrium, Southern Gate, Chichester, West Sussex, PO19 8SQ, United Kingdom.

For details of our global editorial offices, for customer services and for information about
how to apply for permission to reuse the copyright material in this book please see our
website at www.wiley.com/wiley-blackwell.

Library of Congress Cataloging-in-Publication Data

Names: Goodhew, Steve, author.
Title: Sustainable construction processes : a resource text / Steve Goodhew.
Description: Chichester, UK ; Hoboken, NJ : John Wiley & Sons, 2016. |
    Includes bibliographical references and index.
Identifiers: LCCN 2016003739 (print) | LCCN 2016008932 (ebook) | ISBN
    9781405187596 (pbk.) | ISBN 9781119247944 (pdf) | ISBN 9781119247951 (epub)
Subjects: LCSH: Sustainable construction–Great Britain–Textbooks. |
    Sustainable buildings–Great Britain–Design and construction–Textbooks.
Classification: LCC TH880 .G665 2016 (print) | LCC TH880 (ebook) | DDC
    690.068/4–dc23
LC record available at http://lccn.loc.gov/2016003739

A catalogue record for this book is available from the British Library.

Wiley also publishes its books in a variety of electronic formats. Some content that appears
in print may not be available in electronic books.

Cover image: Courtesy of the Author

Set in 10/12.5pt Avenir by SPi Global, Pondicherry, India
Printed and bound in Malaysia by Vivar Printing Sdn Bhd

1 2016

Contents

Preface

This text follows the process of sustainable construction from an idea to the creation of a sustainable building.

When a client or architect imagines a sustainable building, there are a myriad of systems, processes, guidelines, and standards that are negotiated in the journey from blueprint to completed building.

Sustainable Construction Processes: A Resource Text provides a source guide along this journey, whilst also discussing the challenges and uncertainties that arise when constructing a building worthy of its sustainable credentials.

The idea for this book came from the need for a text, suited to Plymouth University's students, that would cover sustainable construction as a process rather than simply focusing on the needs of a designer. It complements but does not duplicate the many other texts on sustainable construction that focus on the design of buildings, procedures, links to sustainable development, climate change, or sustainable cities. Rather, it is concerned with how the best-laid plans of designers, planners, engineers, consultants, and project managers come to fruition, and the process though which buildings are (or are not) constructed in a fashion that makes these plans work in practice. This is a fast-moving field of study, and inevitably different policies, facts, figures, and assessment systems change with a remarkable rapidity. However, the underlying principles of how we build sustainably and the basic tools that are required for this task remain relatively static.

This book introduces the rationale and history that lie behind the drivers for sustainable construction. Decisions inform even the earliest stage of a build, such as the client deciding whether they require a new or refurbished building. From this beginning, the text follows the decision-making process for each stage of a building's life cycle to illuminate the requirements and challenges of designing, constructing, and occupying sustainable buildings. It leads the reader through the areas of sustainable procurement: how we can obtain buildings that will meet the needs of our clients, as well as the broader needs of society and the environment, without costing us and our communities too much. The elements of building design related to energy, water, and materials are examined to demonstrate how construction processes can ensure that our buildings are truly sustainable. Assessment systems are introduced and their basic underlying principles discussed. The challenge of anticipating the behaviours of occupants and the practicalities of building with some of the new sustainable technologies are considered. Finally, we consider potential areas of growth and present some contrasting visions for the future of sustainability.

This definition contains two key concepts:

- the concept of needs, in particular, the essential needs of the world's poor, to which overriding priority should be given, and
- the idea of limitations imposed by technology and social organisation on the environment's ability to meet present and future needs (IISD 2015).

In 1995 the definition was further refined, highlighting three interconnected elements of sustainability:

> Economic development, social development and environmental protection are interdependent and mutually reinforcing components of sustainable development, which is the framework for our efforts to achieve a higher quality of life for all people.
>
> (World Summit on Social Development 1995)

This clarification leads to a concept of sustainability that includes three core components, known as the three E's of sustainable development. These are equity, environment, and economics. So, sustainability can be viewed in the broadest sense as balanced living within the three pillars of sustainable development: economic growth, social progress, and environmental protection. This is also sometimes known as 'the triple bottom line', which could be described as an expanded baseline for measuring performance, adding social and environmental dimensions to the traditional monetary yardstick. Some further development of the triple bottom line also includes governance—how to enact the three pillars—although to a certain extent this is implied in the triple bottom line. A number of other commentators have introduced a fourth pillar, this varying in focus from culture to administration, reminding us that sustainable development has to be culturally appropriate and enacted. (Due to the varying scope of these proposed fourth pillars, here we will stick to three).

Although the three E's model is commonly accepted as the basis for any analysis of sustainability, the Forum for the Future has provided a more penetrating model for analysis of sustainability. Much of this model also takes detail from another system called the Natural Step. More information concerning the Natural Step can be found in Chapter 6.

The model comprises categories of analysis within five broad forms of capital (Forum for the Future 2013):

1. Manufactured capital comprises material goods, or fixed assets that contribute to the production process rather than being the output itself.
2. Natural capital is any stock or flow of energy and material that produces goods and services; this can include carbon sinks that absorb, resources that provide, and processes that maintain.
3. Human capital consists of people's health (both physical and mental), knowledge, skills, and labour—all the things needed for productive work.
4. Social capital concerns the human relationships, partnerships, and institutions that help us maintain and develop human capital in partnership

with others, for example, families, communities, businesses, trade unions, schools, and voluntary organisations.

5. Financial capital is the assets that can be owned and traded, such as shares, bonds, notes, and coin. These play an important role in our economy, enabling other types of capital to be owned and traded.

Using the separate categories described above, the implications of the actions and processes chosen for a construction project or the activities of a construction company can be analysed. However, collective decisions over a period of time can indicate the stance of a company or individual. The reactions of people and organisations in turn can be divided into four very broad leadership or cooperative stances. As they are sometimes used to classify actions or attitudes it is useful to understand their meanings.

1. The Thomist position (coming from the term 'doubting Thomas'), leading from the broad-based doubts concerning any link between environmental problems and our way of life.
2. The 'business as usual' or the 'Macawber syndrome', by which people accept that there are problems with the world's systems but believe that solutions will 'turn up', so they do not need to concern themselves unduly.
3. The 'no-regrets' philosophy. This approach notes the problems with the world's systems and takes the view that concerted action might be helpful as long as it does not 'break the bank'.
4. The precautionary principle. This precautionary principle takes the logical argument that many resources will have to be employed to reduce the impact of unsustainable development, and if in doubt we should take whatever reasonable action is necessary to avoid disaster.

None of these stances have set boundaries, and in some instances people or organisations can exhibit traits that could be ascribed to more than one of these positions.

Now that sustainability has been defined in a general sense, it is appropriate to understand the drivers—environmental, historical, social, and economic—that push governments, companies, and individuals toward a sustainable approach.

1.3.1 Drivers for environmental sustainability

Many drivers for the environmental element of the three pillars exist. Geographically some are close to home: the visible impacts of agricultural land being built upon, or increases in local traffic flow. Others are regional: increased emissions from regional power stations, or the global impact of increasing long-haul air traffic. Often the actions that are taken in everyday life, particularly when constructing a building, can have an impact upon the local, regional, and global stages. The next section of this chapter will introduce some of the science and evidence behind those effects. Figure 1.1 shows a number of events that have triggered or influenced the current economic, technological, social, and environmental situation. It can be

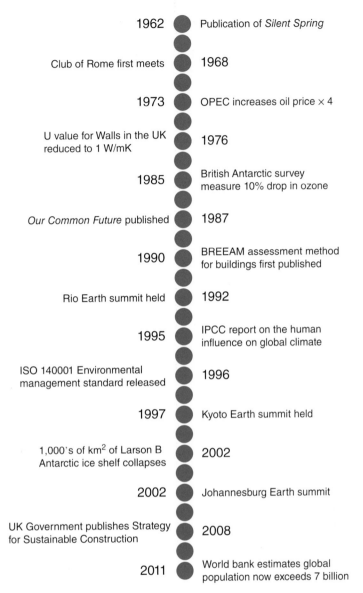

1962	Publication of *Silent Spring*
Club of Rome first meets	1968
1973	OPEC increases oil price × 4
U value for Walls in the UK reduced to 1 W/mK	1976
1985	British Antarctic survey measure 10% drop in ozone
Our Common Future published	1987
1990	BREEAM assessment method for buildings first published
Rio Earth summit held	1992
1995	IPCC report on the human influence on global climate
ISO 140001 Environmental management standard released	1996
1997	Kyoto Earth summit held
1,000's of km² of Larson B Antarctic ice shelf collapses	2002
2002	Johannesburg Earth summit
UK Government publishes Strategy for Sustainable Construction	2008
2011	World bank estimates global population now exceeds 7 billion

Figure 1.1 Events that have triggered or influenced the current economic, technological, social, and environmental situation.

seen that the publication of evidence linking human activity with environmental change has been interspersed with interventions, governance, standards, and agreements to introduce sustainable construction.

One of the most often-quoted drivers for sustainable construction is climate change and the underpinning influence, the greenhouse effect. The 2014 IPCC Climate Change 2014 Synthesis Report states:

Anthropogenic greenhouse gas emissions have increased since the preindustrial era, driven largely by economic and population growth, and are now higher than ever. This has led to atmospheric concentrations of carbon dioxide,

5

methane, and nitrous oxide that are unprecedented in at least the last 800,000 years. Their effects, together with those of other anthropogenic drivers, have been detected throughout the climate system and are extremely likely to have been the dominant cause of the observed warming since the mid-20th century (IPCC 2014).

It is therefore logical to look at the wider context and science behind this effect.

1.3.2   Climate change

The temperature of interstellar space is approximately −250°C, whereas the range of surface temperatures of the Earth is between −25°C and 45°C (NASA 2000). The Earth orbits the Sun, a large source of many wavelengths of electromagnetic radiation (including infrared 'radiant' heat); as radiant heat hits any surface, such as the surface of the Earth, depending on the surface's characteristics, some of that radiant heat will be absorbed, raising the temperature on and around that surface. The closer the surface is to the source of radiant heat, the more infrared energy will be absorbed. This influence can account for most of the difference in temperature between space and the Earth's surface, but the Moon (a body approximately the same distance from the Sun, albeit smaller in mass) can experience temperatures lower than −100°C on its dark side (more details in Text Box 1.1). The main physical attributes that differentiate the Earth from other planets in the solar system are related to its atmospheric gases, which account for the less extreme close-to-surface temperature variations.

Water vapour (H_2O), carbon dioxide (CO_2), and methane (CH_4), which are the most important and often naturally generated greenhouse gases, are transparent to short wavelength radiation (produced at high temperatures from the Sun) but opaque to longer wavelengths of infrared radiation (heat radiation from the relatively cooler Earth's surface; see Figure 1.2).

The net result is that radiant heat from the Sun is allowed through our atmosphere to warm the Earth's surface, but the infrared radiation of a longer wavelength emitted from the Earth's surface accounts for the extra 30°C on our planet's surface.

The 'greenhouse effect' is perfectly natural; however, concern is centred on the rapid increase in greenhouse concentrations due to man's (anthropogenic) activities over and above those that are naturally present.

The principal greenhouse gases to which man contributes are carbon dioxide (CO_2; 57% fossil fuel emissions; 17% deforestation, decay of biomass; 3% other), chlorofluorocarbons (CFCs; 1%, assuming the Montreal protocol maintains its effectiveness), methane (CH_4; 14%), agricultural activities, and waste management (IPCC 2007). Other types of emissions include nitrous oxide (4%, often associated with vehicle emissions), ozone (O_3), and black carbon (BC are particulates of carbon, not a gas that can contribute to atmospheric warming) (Table 1.1).

1.1   Want to know more? What might be the effect upon Earth in the future? Why are greenhouse gases so important?

The Earth, the third planet from the sun in our solar system, is surrounded by a gaseous atmosphere and supports life. To understand our environment fully, it is helpful to build a picture of the planet's physical position in space.

The Solar System
Approximate average distance from Sun (varies according to orbital position)

Mercury	60 million km (37 million miles)
Venus	110 million km (70 million miles)
Earth	150 million km (93 million miles)
Mars	225 million km (140 million miles)
Jupiter	800 million km (500 million miles)
Saturn	1,430 million km (900 million miles)
Uranus	2,900 million km (1,800 million miles)
Neptune	4,500 million km (2,800 million miles)
Pluto	5,900 million km (3,666 million miles)

To take a considered peek into the future of our planet with increased quantities of greenhouse gases it is logical to look at those planets that also orbit our Sun and see if there are any lessons that can be learned from their surface conditions in relation to the constituents of their atmosphere. As can be seen to the left, the known planets in our solar system extend from Mercury, very close to the Sun in solar distance terms, to Pluto, which is very distant. Naturally, the planets that are closer to an intense heat source (the Sun) have higher surface temperatures, but if we analyse the impact that atmospheres can have, there are some interesting estimates of surface temperature readings. It would be wise to compare the Earth with its nearest two neighbouring planets, both closer and farther from the Sun. Whilst Earth is a little larger than Venus and Mars, all three are similar enough in size and have orbits that are relatively close to each other to bear a valid comparison. Whilst Venus is the closest to the Sun of the three, the surface temperature of 450°C is high. Part of the explanation lies in the composition of the Venution atmosphere, which is composed mainly of carbon dioxide (96% and around 300 times the amount of CO_2 as Mars (NASA 2015a)). Much of the solar energy it receives is trapped by this atmosphere, resulting in high estimates of temperatures both on the sides of the planet facing to and away from the Sun (Goldsmith 1990). Mars has a thin atmosphere, again mainly carbon dioxide (96.5%) with an average surface temperature of about −50°C at night and about 0°C during a summer's day. The Martian poles can witness extremes of −153°C (NASA 2014).

The Earth's surface temperatures vary from −89°C to 57°C (Cain 2008a). So, if the Earth's atmosphere had low concentrations of atmospheric gases, then the temperature on Earth could be more like the Moon, which rises to as much as 116°C in the day and then dips down to as low as −173°C at night. Alternatively, with much more carbon dioxide and other greenhouse gases, it could be more similar to Venus (Cain 2008b).

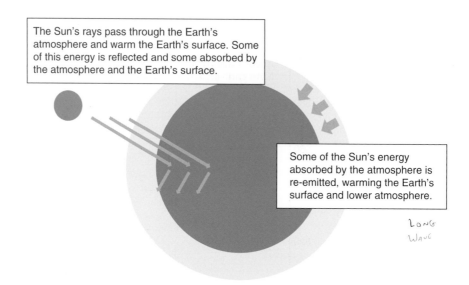

The Sun's rays pass through the Earth's atmosphere and warm the Earth's surface. Some of this energy is reflected and some absorbed by the atmosphere and the Earth's surface.

Some of the Sun's energy absorbed by the atmosphere is re-emitted, warming the Earth's surface and lower atmosphere.

Long Wave

Figure 1.2 Greenhouse gases in the Earth's atmosphere absorb heat and warm the Earth's surface to a temperature that will support life, concepts based on information contained in NASA (2014).

Table 1.1 The global warming potential (GWP) of a small number of the more commonly quoted greenhouse gases (Data taken from Table 2.14 in the IPCC document *Climate Change 2007*, Working Group 1)

Substance	Chemical Formula	Atmospheric Lifetime (yrs)	GWP 20 years	GWP 100 years	GWP 500 years
Carbon dioxide	CO_2	Varies	1	1	1
Methane	CH_4	Varies but approx. 12	72	25	7.6
Nitrous oxide	N_2O	114	289	298	153
CFC-12 (controlled by the Montreal Protocol)	CCl_3F_3	100	11,000	10,900	5,200
HFC-32	CH_2F_2	4.9	2,330	675	205

The degree to which any particular greenhouse gas affects global warming depends on two factors:

- Its relative effectiveness (per unit of concentration) in blocking that low temperature radiation from the Earth
- Its concentration in the Earth's atmosphere

In some quarters a debate continues surrounding the link between human activities, the previously mentioned emissions of greenhouse gases, and the impact upon the climate. James Lovelock (Lovelock 2000) points out that the present chemical composition of the Earth's atmosphere is 'highly improbable', given the expectations of orthodox chemistry. It contains a mixture of gases that should react with each other so that,

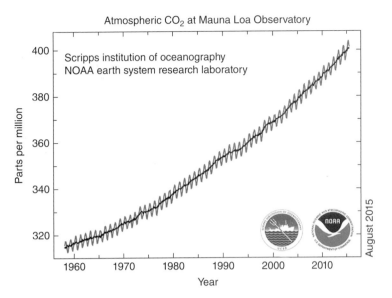

Figure 1.3 Atmospheric carbon dioxide measurements at Mauna Loa Observatory, Hawaii (NOAA 2015).

in chemical equilibrium, only traces of the original gases remain. For example, if oxygen is present together with methane, a reaction could take place that would use the oxygen in the presence of sunlight to form carbon dioxide and water. Therefore, without new sources of methane and oxygen to replace these gases, they would eventually become exhausted, as is probably the case for Venus and Mars (see Text Box 1.1). Thus, according to Lovelock, it is very likely that living organisms through their bodily processes and activities keep the Earth's atmosphere in the unique state that permits life to survive (The Gaia hypothesis). If living organisms can be linked to the development of the concentration of gases in the past, it is not illogical to link the present activities of living organisms to the same changes.

Of all the greenhouses gases in the Earth's atmosphere only carbon dioxide has a global warming potential of 'one' but is one of those most associated with human activities. Through measurements made since the 1950s at an observatory in Mauna Loa, Hawaii, an increasing trend of CO_2 atmospheric concentrations can be seen (Figure 1.3).

Hawaii, whilst being in the northern (most industrialised) hemisphere, is also a long way from any sources of CO_2 that could have an impact upon these measurements. They can therefore be assumed to be a reasonably true reflection of how CO_2 atmospheric concentrations are increasing globally. Figure 1.4 (IPCC 2007) shows global temperature observations for land and oceans alongside simulated temperature predictions, with both natural forcings and natural and anthropogenic forcings. It is quite evident that the upward path of the global temperature observations follows the approximate path of the simulation that includes emissions from human-derived sources.

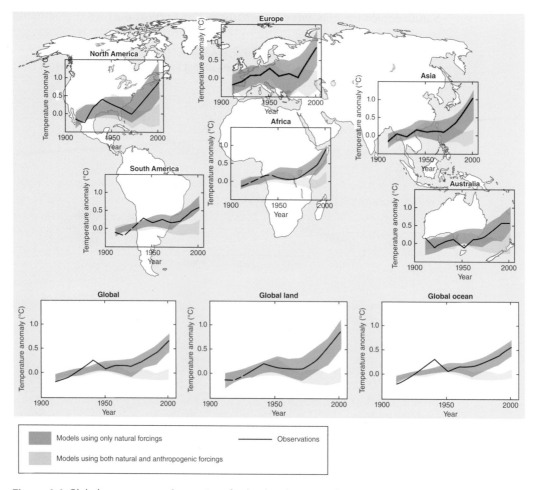

Figure 1.4 Global temperature observations for land and oceans alongside simulated temperature predictions both with only natural forcings (blue shading) and natural and anthropogenic forcings (dark blue shading).

1.3.3 Historical influences on the environment

So if the evidence points to a link between recent CO_2 and global temperature rise, how much of this may be attributed to the activities of mankind? When did this influence start? Various different methods of recording climate and temperatures alongside older atmospheric CO_2 measurements have resulted in a pattern of relatively stable CO_2 measurements until the end of the 18th and beginning of the 19th centuries. The timing of these indicators and much of the influence upon the global environment can be traced back to two transformations: the UK's agricultural and industrial revolutions. The first of these two followed the enclosure of much of the UK's commonly owned land and the development of new agricultural technologies. In the medieval period farming in the British Isles was undertaken in open fields, where tenant farmers tended small strips of land. From Tudor times these separate strips of land were merged into individually owned or rented fields. From 1750 onwards the enclosure of the land by parliamentary act was common. In the period between 1604 and 1914 over

Earth in the Balance (Gore 1992) describes environmental issues from a global point of view and, further, outlines a range of policies that would tackle the most important of those issues. Written by a US vice president, this volume had some influence and good sales whilst also having some negative press. Many of the predictions concerning climate change, habitat loss, water quality, and so forth that were described in this text have proven to be accurate. *Cradle to Cradle: Remaking the Way We Make Things*, by William McDonough and Michael Braungart (2002) calls for a radical change in the industrial pattern of making, using, and disposing of things. The authors propose a change from a cradle-to-grave use pattern to a cradle-to-cradle focus, proposing the manufacture of products that can be and have been upcycled. The use of a lifecycle development philosophy recommends that when products have reached the end of their useful life, they either degrade to 'biological nutrients' or are used again as 'technical nutrients'. The collective power of these texts and scientific reports such as those from the previously mentioned IPCC, reinforced by images such as 'Earthrise', taken of the Earth from the Moon (NASA 2015a), have led to a feeling that our planet and its resources are precious. To quantify this preciousness of the Earth and its relationship to human activities, researchers have introduced an assessment referred to as ecological footprinting.

1.3.5   Ecological footprinting

One of the major drivers for sustainable construction is the need for people, countries, and continents to be able to build, renovate, and operate buildings within the available supply of natural resources (often termed 'natural capital'; for further detail, see section 1.2). Factors affecting the supply of natural capital include the more apparent depletion of finite fossil fuels or the less easily quantified areas of land needed to deal with waste from human activities. An original study undertaken in 1997 focused upon the relationship between natural capital and the consumption and activities of most of the earth's human population (Wackernagel *et al.* 1997). This study laid the foundations for the later, more detailed, footprinting methodologies and reviewed data from the 52 nations that represent 80% of the global population and generate 95% of the worldwide domestic product. Mathis Wackernagel, one of the first proponents of establishing a tool for measuring our need for natural capital, describes ecological footprinting as 'an accounting tool that can aggregate ecological consumption in an ecologically meaningful way. It gives us, therefore, a realistic picture of where we are in ecological terms' (Wackernagel *et al.* 1999, p. 389, section 5 conclusions).

The methods used to calculate ecological footprints are relatively straightforward. The activities/consumption of the population of each nation are estimated in terms of areas of agricultural or biological production and area needed to assimilate any wastes generated. This process is aided by the readily available data (most from UN or national sources) that describe ecological productivity, resource production, and trade.

The Global Footprint Network has the most current data for 2010 (published in 2013). Ecological footprints can also be combined with other measurements or assessments such as the Human Development Index (HDI) to extend the ability to analyse the performance of nations or regions. The HDI is a summary measure of human development and it measures the average achievements in a country in three basic dimensions of human development:

- A long and healthy life, as measured by life expectancy at birth
- Knowledge, as measured by the adult literacy rate (with two-thirds weight) and the combined primary, secondary, and tertiary gross enrolment ratio (with one-third weight)
- A decent standard of living, as measured by GDP per capita (PPP US$) (UNDP 2004)

Table 1.2 Human Development Index (HDI) and ecological footprint of selected countries of varying incomes (Moran *et al.* 2008; Global Footprint Network 2010)

Country	HDI	Ecological Footprint of Consumption (gha per person)[a]	Footprint to Global Biocapacity Ratio[b]
Norway[c]	0.96	5.9	3.2
UAE	0.85	11.9	6.5
Panama	0.80	1.9	1.0
India	0.60	0.8	0.4
Bangaladesh[d]	0.52	0.5	0.5
Niger	0.28	1.1	0.6

[a] gha = global hectares. This is a measure of how much land and sea are needed to supply all the resources consumed by one individual.
[b] This ratio shows how much larger the per capita demand on resources is compared to the per capita biocapacity available worldwide. It represents the number of planet Earths that would be required to support the current population at that country's level of consumption (assuming no biological productivity is reserved for the use of wild species).
[c] Highest and lowest HDI score of reported countries.
[d] Highest and lowest ecological footprint per capita of reported countries.

Table 1.3 Ecological footprint and biological capacity (all data from public domain figures)

Level of average income per capita	Population (million)	Ecological Footprint of Consumption (gha per person)[a]	Total Biocapacity (gha per person)	Ecological Deficit
High income countries	1031.4	6.1	3.1	−3.0
Middle income countries	4323.3	2.0	1.7	−0.2
Low income countries	1303.3	1.2	1.1	−0.1
World	6671.6	2.7	1.8	−0.9

[a] gha = global hectares. This is a measure of how much land and sea are needed to supply one human with the resources he or she consumes.
Source: Global Footprint Network (2010).

Moran *et al.* (2008) show the HDI and ecological footprint of selected countries using 2003 data, and this is described in Table 1.2. The HDI column shows a gradual numerical reduction in the human development index alongside the corresponding ecological footprint. The final column shows a deficit in relation to ecological area. This is the discrepancy between consumption and capacity for the whole Earth. This deficit is not evenly shared between the different development levels of the Earth's countries. The deficit includes high development/income countries, such as Norway, which on average have a footprint to global biocapacity ratio of three planets and low development/income countries such as Niger with a 0.6 footprint to global biocapacity ratio. These figures help communicate a stark message: as populations rise and development across the globe puts more pressure upon the global natural capital, we are increasingly living beyond the ability of our planet to supply our needs. Whilst there is no single solution, sustainable development, including sustainable construction, is a large part of any resolution to these problems.

Other extensions of footprinting directed at construction materials and processes are developing at a rapid rate, and some of these are discussed in Chapter 6, which is devoted to assessment systems (Gilroy-Scott *et al.* 2013). One of the debates associated with reducing ecological footprints is related to our behaviour and choices and how much these might cost, both now and in the future.

1.3.6 Economic drivers for actions to reduce the impact of climate change

One of the more effective methods of evaluating the connections between behaviour and the decisions people make is to link the environmental consequences of an action to economic impact. This is by no means a perfect linkage; something that might make economic sense might not be an effective solution in many other ways. However, when longer term economic impacts are linked to environmental and social impacts—and these less basic monetary issues are given a price tag—the gravity of some situations can be made more apparent. This was highlighted by a speech by the governor of the Bank of England, Mark Carney, who in September 2015 stated, 'The far-sighted amongst you are anticipating broader global impacts on property, migration, and political stability, as well as food and water security' (BBC 2015). Carbon taxes and issues related to sequestration (set-aside farming land) could all be seen to be economic levers that authorities or governments are using to alter behaviour in relation to the more environmental and social aspects of behaviour. They are, however, end-of-stream monetary interventions. An overview of the linkages between the environment and economics was provided by the *Stern Review* (Stern 2006).

The Stern report came from an independent review, commissioned by the UK's Chancellor of the Exchequer and reported in 2006. The review examines the evidence related to the economic impacts of climate change, explores the economics of stabilising greenhouse gases in the atmosphere through an international perspective, and considers policy challenges to manage the transition to a low-carbon economy and climate change adaption.

The review establishes that long-term greenhouse-gas emissions need to be cut to a level of 550 ppm CO_2e (see Chapter 3 for an explanation of CO_2e as opposed to CO_2) in four ways and acknowledges that the costs associated with each will differ considerably depending on which combination of the following methods is used, and in which sector:

1. Reducing demand for emissions-intensive goods and services
2. Increased efficiency, which can save both money and emissions
3. Action on non-energy emissions, such as avoiding deforestation
4. Switching to lower-carbon technologies for power, heat, and transport

Construction and buildings can play a significant part in all: through their design, their specification, their construction, their use, and their removal.

The conclusions of the report were many and complex, but risking a sound-bite approach, one quotation is especially impactful:

> Resource cost estimates suggest that an upper bound for the expected annual cost of emissions reductions consistent with a trajectory leading to stabilisation at 550 ppm CO_2e is likely to be around 1% of GDP by 2050.
>
> (Stern 2006)

One percent of GDP is obviously a variable figure, implying a larger cost to more wealthy countries (as defined by GDP) than less wealthy countries. For any country this figure will represent a considerable expenditure, but when judged against the whole wealth of a country, it can be considered modest. Since the review was completed, the key measured environmental indicators have provided a better view of the trends associated with the state of the environment. This has understandably changed and caused Sir Nicholas Stern to reflect upon the 1% figure:

> Looking back, I underestimated the risks. The planet and the atmosphere seem to be absorbing less carbon than we expected, and emissions are rising pretty strongly. Some of the effects are coming through more quickly than we thought then.
>
> (Stewart & Elliott 2013)

This statement is amplified by a further, more pointed, comment:

> This is potentially so dangerous that we have to act strongly. Do we want to play Russian roulette with two bullets or one? These risks for many people are existential.
>
> (Stewart & Elliott 2013)

The economic case can therefore be stated as being considerable, and through the judgment of the Stern Review, environmental, and by inference, socially related expenditure makes good sense.

1.4   The environmental importance of design, construction, and care of buildings

According to the Organisation for Economic Co-operation and Development (OECD), buildings have levels of consumption and impact upon waste that focus attention upon most aspects of their design, construction, and after-care (Hartenberger 2011). These impacts can all be linked to the result of decisions made by building professionals. Those decisions will be influenced by the traditional needs of the construction industry, the restrictions imposed by legislation, and guidance and the needs of the clients for those buildings and projects. How to mesh the need to reduce the impacts of the industry in the widest sense and remain true to those traditional client needs is one of the main drivers of this book.

1.5   Where next?

It can be seen that whilst there are uncertainties associated with the extent of anthropogenic environmental impacts, it is clear that a sustainable course needs to be charted. This will involve a number of changes to the way that we design, build, maintain, and run our buildings. The following chapters take the reader through procurement of a sustainable building and the contractual issues linked to this. The design-related issues are then discussed with a general split in focus between energy and materials. The construction process is examined, discussing the importance of actions taken at the different stages of the building process. Assessment systems and the part they play in encouraging and maintaining compliance of the performance of buildings is described and debated. The actual use of sustainable technologies is assessed, along with the ways in which designers, clients, contractors, and facility managers can ensure those systems work to their advantage. Finally, the future aspects of sustainable construction are examined, and an attempt is made at the dangerous but fun task of guessing in an informed way what the construction industry might be most employed in thinking about most in 25 years' time (or less).

References

BBC (2015) Bank of England's Carney warns of climate change risk [Online]. Available: http://www.bbc.co.uk/news/business-34396961 [30 Sept 2015].

Cain, F. (2008a) Temperature of the moon [Online]. Available: http://www.universetoday.com/19623/temperature-of-the-moon/ [12 March 2015].

Cain, F. (2008b) Venus greenhouse effect, [Online], Available: http://www.universetoday.com/22577/venus-greenhouse-effect/ [12th March 2015].

Carson, R. (1951) *The sea around us*, Oxford University Press, Oxford, 1991.

Carson, R. (1962) *Silent spring*, Mariner Books, New York, 2002.

Forum for the Future (2013) Five capitals model [Online]. Available: http://www.forumforthefuture.org/sites/default/files/project/downloads/five-capitals-model.pdf [16 Dec 2013].

Gilroy-Scott, B., Chilton, J., & Goodhew, S. (2013) Earth footprint of the construction phase of the Wales Institute for Sustainable Education at the Centre for Alternative Technology, *International Journal of Sustainability Education* **8**(2), 73–91.

Global Footprint Network (2010) The 2007 data tables [Online]. Available: http://www.footprintnetwork.org/images/uploads/NFA_2010_Results.xls [16 Dec 2013].

Goldsmith, E. (1990) *5000 days to save the planet*, Hamlyn, London.

Gore, A. (1992) *Earth in the balance: Ecology and the human spirit*, Rodale, Emmaus, PN, 2006.

Hartenberger, U. (2011) Why buildings matter [Online], 1 Jul 2011. Available: http://www.theguardian.com/sustainable-business/sustainable-building [16 Dec 2013].

Hodgson, G. M. (2004) *Biographical dictionary of British economists* (ed D. Rutherford), Thoemmes Continuum, Bristol.

IISD (2015) International Institute for Sustainable Development [Online]. Available: https://www.iisd.org/sd/ [31 August 2015].

IPCC (2007) *Climate change 2007: Synthesis report. Contribution of Working Groups I, II and III to the fourth assessment report of the Intergovernmental Panel on Climate Change* (eds R.K. Pachauri and A. Reisinger). IPCC, Geneva, Switzerland, 104 pp. [Online]. Available: https://www.ipcc.ch/publications_and_data/ar4/syr/en/spms2.html# [16 Dec 2013].

IPCC (2014) *Climate Change 2014: Synthesis Report. Contribution of Working Groups I, II and III to the Fifth Assessment Report of the Intergovernmental Panel on Climate Change* (R.K. Pachauri and L.A. Meyer, eds.), IPCC, Geneva, Switzerland.

Leopold A. (1949) *A Sand County almanac*, Ballantine Books, New York, 1986.

Lovelock, J. E. (1979) *Gaia: A new look at life on Earth*, Oxford University Press, Oxford.

Lovelock, J. E. (2000) *Ages of Gaia* (2nd ed reprint with new preface), Oxford University Press, Oxford.

Lovelock, J. E. (2007) *The revenge of Gaia: Why the Earth is fighting back and how we can still save humanity*, Penguin, London.

Malthus, T. R. (1798) *An essay on the principle of population, as it affects the future improvement of society, with remarks on the speculations of Mr. Godwin, M. Condorcet, and other writers*, J. Johnson, London.

McDonough, W., & Braungart, M. (2002) *Cradle to cradle: Remaking the way we make things*, North Point Press, New York.

Meadows, D. H., Meadows, D. L., Randers, J., & Behrens, W. W. III (1972) *Limits to growth*, Signet, London.

Molina, M. J., & Rowland, F. S. (1974) Stratospheric sink for chlorofluoromethanes: Chlorine atomcatalysed destruction of ozone, *Nature* **249**(5460), 810–812.

Mollison, B. (1978) *Permaculture: A designer's manual*, Tagari Publications, Tasmania, Australia.

Moran, D., Wackernagel, M., Kitzes, J., Goldfinger, S., & Boutaud, A. (2008) Measuring sustainable development: Nation by nation, *Ecological Economics* **64**(3), 470–474, ISSN 0921-8009 [Online]. Available: 10.1016/j.ecolecon.2007.08.017 [31 August 2015].

NASA (2000) NASA, Earth Observatory, land surface temperature [Online]. Available: http://earthobservatory.nasa.gov/GlobalMaps/view.php?d1=MOD11C1_M_LSTDA [7 Oct 2015].

NASA (2014) U.S. Department of Commerce National Oceanic and Atmospheric Administration NOAA Research [Online]. Available: http://www.esrl.noaa.gov/gmd/ [20 June 2014].

NASA (2015a) Global climate change: Vital signs of the planet [Online]. Available: http://climate.nasa.gov/causes/ [2 June 2015].

NASA (2015b) Earthrise [Online]. Available: http://www.nasa.gov/multimedia/imagegallery/image_feature_1249.html [7 Oct 2015].

NOAA (2015) U.S. Department of Commerce National Oceanic and Atmospheric Administration NOAA Research [Online]. Available: http://www.esrl.noaa.gov/gmd/webdata/ccgg/trends/co2_data_mlo.png [30 August 2015].

Oxford Dictionaries (2014) Sustainability [Online]. Available: http://www.oxforddictionaries.com/ [20 June 2014].

Papanek, V. (1971) *Design for the Real World: Human Ecology and Social Change*, Pantheon Books, New York.

Reisner, M. (1986) *Cadillac desert: The American west and its disappearing water*, Penguin, London, 1993.

Ruskin, J. (1862) *Unto this last*, M. Dent and Sons Ltd, London and Toronto, 1921 edition available Library of Congress [Online]. Available: https://archive.org/details/untothislast00rusk [19 June 2015].

Schumacher, E. F. (1973) *Small is beautiful: Economics as if people mattered*, Harper Perennial, New York, 1989.

Stern, N. (2006) *Stern review on the economics of climate change*, Cambridge University Press [Online]. Available: http://webarchive.nationalarchives.gov.uk/+/http:/www.hm-treasury.gov.uk/sternreview_index.htm [16 Dec 2013].

Stewart, H., & Elliott, L. (2013) Nicholas Stern: 'I got it wrong on climate change—it's far, far worse', *The Guardian*, 26 Jan 2013 [Online]. Available: http://www.theguardian.com/environment/2013/jan/27/nicholas-stern-climate-change-davos [16 Dec 2013].

UK Parliament (2015) Managing and owning the landscape [Online]. Available: http://www.parliament.uk/about/living-heritage/transformingsociety/towncountry/landscape/overview/enclosingland/ [20 June 2014].

UNDP (2004) United Nations Development Programme page 268, Note on statistics in the human development report [Online]. Available: http://hdr.undp.org/reports/global/2004/pdf/hdr04_backmatter_2.pdf [31 August 2015].

Wackernagel, M., Onisto, L., Bello, P., *et al.* (1999) National natural capital accounting with the ecological footprint concept, *Ecological Economics* **29**, 375–3.

Wackernagel, M., Onisto, L., Callejas L., *et al.* (1997) *Ecological footprints of nations: How much nature do they use? How much nature do they have?* Commissioned by the Earth Council for the Rio5 Forum, International Council for Local Environmental Initiatives, Toronto.

Wall, D. (1994) *Green history: A reader*, Routledge, London.

World Bank (2015) World Bank data section [Online]. Available: http://data.worldbank.org/indicator/SP.POP.TOTL/countries?display=graph [31 August 2015].

World Commission on Environment and Development (1987) *Our common future*, Oxford University Press, Oxford.

World Summit on Social Development (1995) *Copenhagen declaration on social development*, United Nations, New York [Online]. Available: http://www.un.org/documents/ga/conf166/aconf166-9.htm [16 Dec 2013].

2. Procurement and sustainability

2.1 Procurement and construction

What do we mean by procurement? This term normally describes the act of buying or obtaining something (a product, a space, a service) on behalf of someone else. This can work on a small scale where a tradesperson can supply and fit materials and products for a householder, where the householder might have considerable input into the act of procuring the 'right' product or materials for them. Most large organisations that work within the sphere of construction will have dedicated parts of the organisation that focus solely on procuring the right product, service, or material of the appropriate quality and for the best price. For some, more specialist, construction projects, this might involve suppliers being nominated to service a particular building or series of buildings. Buyers may be able to negotiate more favorable deals when buying in quantity or over a sustained period.

Procurement related to construction has a more finely tuned meaning. The Joint Contracts Tribunal defines procurement related to construction as 'the activities undertaken by a client or employer who is seeking to bring about the construction or refurbishment of a building' (JCT 2013c). In this instance, construction procurement is directly linked to how, in a strategic sense, contracts for construction work are produced, managed, and completed. Because of this more contractually strategic meaning, procurement takes on more importance than merely purchasing items, materials, and services. This importance extends to funding and organising construction at all stages of development (Hughes *et al*. 2006).

2.1.1 Why the way in which we procure our buildings impacts upon sustainability

The procedures and processes through which we procure our buildings and structures can have repercussions for the sustainability of a building. The performance of a building, its immediate environment, and the way it has

Sustainable Construction Processes: A Resource Text, First Edition. Steve Goodhew.
© 2016 John Wiley & Sons, Ltd. Published 2016 by John Wiley & Sons, Ltd.

been constructed and looked after once it has been handed over to the client all influence whether it can be described as a success in terms of sustainability. If these aspects are part of a client's priorities, then how shall these be defined as objectives within the procurement process? If the client is ambivalent, should these aspects be given equal billing against time, cost, and quality? Or do these in themselves encompass the means to the end of procuring more sustainable buildings? We need to establish the mechanisms for obtaining our built environment before investigating the details of the sustainability of the product and process. Public and private sector procurement necessarily differ in approach, but the sustainability of the process, product, and after care must be equally balanced for either sector, even if the means of achievement are different.

Obtaining a building, either for a client or as a sole purchaser, is markedly different from procuring any other commodity, either manufactured or a natural resource. The difference lies in the nature of what buildings are and how they are produced. No two buildings are completely alike; their use, client, and location will normally differ. Thus, a traditional manufacturing process (market and consumer research, producing a brief, a prototype, testing to modify the prototype before full-scale production is undertaken) is not applicable to the production of the majority of buildings. Once assessed through the prototype, the sustainability of a final manufactured product can be refined and repeated. Not so a building when procured through the methods prevalent in the construction industry. As far back as 1962 UK government reports identified the separation of design from production as a problem (Emmerson 1962), and this theme was also picked up by Sir Michael Latham in his 1994 report (Latham 1994). Clients tend not to be bothered as to the source of any issues discovered when accepting a finished project. Their preoccupation tends to be with the fitness for purpose, perceived quality, appearance, and general performance of the final product, not the processes that led to the manufacture of that product. As buildings and structures are products that tend to have a long life, the sustainability issues related to a building's birth and end of life are profound, and the attendant influence upon how and what should be constructed is lasting. In some instances buildings/structures are practically an investment good, giving rise to separate issues connected with the main purpose of a building as a product. This series of complexities and the different influences that a building has upon it emphasises the importance of embedding sustainability through the procurement route as one of the methods of influencing the design, construction, use, reuse, and final solutions for the building.

2.2   Drivers for and concepts behind sustainable procurement

Traditionally the main aim of most construction procurement processes is to obtain best value for a client when obtaining a building or structure: 'Client satisfaction is increasingly seen by all concerned with the development and construction process to be largely dependent upon the selection of the most appropriate procurement methodology' (Morledge & Smith 2013).

This focus on best value and client satisfaction is relatively narrow in scope and mainly views the act of obtaining any building or structure as a form of capital expenditure taking place at one point in time. This approach could be seen as less concerned with the longer-term continuous expenditure associated with the use of the building post-handover and more focused on the building's ability to function 'here and now'. This longer-term, more sustainable approach and its impact are described in a number of texts, including Morledge and Smith (2013), and can be seen as a further development of client expectations.

Guidance concerning procurement from the UK's Commission for Architecture and the Built Environment (CABE) focuses upon the design-related issues and is articulated in a procurement strategy (CABE 2009). This very useful text aligns procurement with local authority work and describes the many standards that can steer the detail of a brief, such as 'Building for Life, Housing Quality Indicators (HQI), Lifetime Homes, Secured by Design and the Code for Sustainable Homes'. It also gives a number of case studies, some of which follow the requirements of the OJEC (*Official Journal of the European Community*). This is the publication in which, according to EU legislation, all European tenders from the public sector that are valued above a certain financial threshold must be published. One of the latter case studies based in Sheffield, UK, uses environmental sustainability as one of the seven criteria. As well as achieving the required environmental standards, developers are asked for their proposals to support more sustainable lifestyles (CABE 2009).

A number of UK organisations have been central in moving the agenda towards a longer-term and wider view by contributing the principles and processes that should underpin future procurement of buildings. A document titled 'Achieving Sustainability in Construction Procurement', produced by the Sustainability Action Group of the Government Construction Clients' Panel (GCCP 2000), outlined a series of principles for government procurement in construction-related activities that would result in the following:

- Procurement in line with value-for-money principles on the basis of 'whole life costs'
- Less waste during construction and in operation
- Targets for energy and water consumption for new projects that meet at least current best practice for construction type and which contribute significantly to the achievement of cross-government targets agreed by Ministers
- The protection of habitat and species, taking due account of the UK Biodiversity Action Plan and the biodiversity action checklist for departments agreed by Green Ministers
- Targets developed in terms of 'respect for people' for procurement of the government estate
- A contribution to the goals of less pollution, better environmental management, and improved health and safety on construction sites

The GCCP document (GCCP 2000) further recommends themes for action and advises a number of possible actions that can be taken forward as ways of progressively meeting the ideals suggested by each theme (see Table 2.1). While these stray from the more tightly defined areas associated with procurement that can be termed sustainable, they clearly show how thinking in this field has moved on since 2000.

In 2006 Sir Neville Sims wrote in the foreword to a document from DEFRA (Sims 2006) that sustainable procurement was 'using procurement to support wider social, economic and environmental objectives in ways that offer long-term benefits'.

Table 2.1 Showing the themes for action that relate to the GCCP document *Achieving Sustainability in Construction Procurement*

Theme for Action	Details of Possible Actions
Reuse existing built assets	Consider the need for new build. Refurbishment/reuse may work better. Think brownfield wherever possible for new construction.
Design for minimum waste	Design out waste both during construction and from the useful life and afterlife of the building or structure. Think whole life costs. Involve the supply chain. Specify performance requirements with care to encourage more efficient use of resources. Think about using recycled materials.
Aim for lean construction	Work on continuous improvement, waste elimination, strong user focus, value for money, high-quality management of projects and supply chains, improved communications.
Minimise energy in construction	Be aware of the energy consumed in the production and transport of construction products. Adopt 'green' travel policies.
Minimise energy in use	Consider more energy efficient solutions in design, including passive systems using natural light, air movement, and thermal mass, as well as solutions involving energy produced from renewable sources.
Do not pollute	Understand your environmental impacts and have policies and systems to manage them positively. Use environmental management systems under ISO 14001 or EMAS. Specify adoption of the Considerate Constructors Scheme 14 or similar.
Preserve and enhance bio-diversity	Look for opportunities throughout the construction process from the extraction of raw materials, through the construction phase, to the landscaping of buildings and estates to provide and protect habitats.
Conserve water resources	Design for increased water efficiency in building services and water conservation within the built environment.
Respect people and their local environment	Be responsive to the community in planning and undertaking construction. Consider all those who have an interest in the project (employees, the local community, contractors).
Set targets	Measure and compare your performance with others. Set targets for continuous improvement. Develop appropriate management systems.

This action plan also continued to adhere to the theory (declared in the 2000 GCCP document described previously) that if a policy of sustainable procurement underpinned the way that government spent its resources then this would, by implication, also impact the way UK companies and supply chains did business. The guiding principles that direct this theme also coincide with the general thinking behind the introduction of building information modelling that is required for UK government construction projects from 2016 (more detail on building information modelling and its uses are discussed in Chapters 5 and 6).

A more recent focus (OGC 2007) has been upon the whole life of the service or facility, particularly when applied to public sector or governmental procurement. This focus has been reiterated by the UK Government's Strategy for Sustainable Construction (HMS Government and Strategic Forum for Construction 2008), in which one of the main drivers for focusing upon procurement is the potential annual saving to the public purse of £2.6 billion, estimated by the 2005 National Audit Office report *Improving Public Services through Better Construction*. The strategy indicates that this potential cost saving is achieved by encouraging the adoption of a number of construction commitments, with a focus upon improved whole life value through the promotion of best practice construction procurement and supply side integration. The detailed construction commitments can be found at www.strategicforum.org.uk, and a summary of these is included in Table 2.2.

Table 2.2 Construction commitments (strategic forum 2012)

Construction Commitments	
Procurement and integration	A successful procurement policy requires ethical sourcing, enables best value to be achieved, and encourages the early involvement of the supply chain. An integrated project team works together to achieve the best possible solution in terms of design, buildability, environmental performance, and sustainable development.
Commitment to people	Valuing people leads to a more productive and engaged workforce, facilitates recruitment and retention of staff, and engages local communities positively in construction projects.
Client leadership	Client leadership is vital to the success of any project and enables the construction industry to perform at its best.
Sustainability	Sustainability lies at the heart of design and construction. A sustainable approach will bring full and lasting environmental, social, and economic benefits.
Design quality	The design should be creative, imaginative, sustainable, and capable of meeting delivery objectives. Quality in design and construction utilising the best of modern methods will ensure that the project meets the needs of all stakeholders, both functionally and architecturally.
Health and safety	Health and safety is integral to the success of any project, from design and construction to subsequent operation and maintenance.

The adoption of these principles is both laudable and sensible, but sustainability is not their main focus, as indicated by the statement with which they are prefaced: 'Delivering construction projects on time, safely and to budget'. A further generic (in this case, not concentrated upon the requirements of the construction industry) but nevertheless very influential and useful document is the British standard for sustainable procurement (BS 8903) (BSI 2010). Action Sustainability were the technical authors of BS 8903 and also provide a range of free sustainable procurement workshops and online tools to assess against BS 8903 (Action Sustainability 2012a). BS 8903 gives any users of the standard a common approach to sustainable procurement, and whilst the Strategic Forum for Construction, the National Audit Office, and the Office of Government Commerce have, as previously stated, pointed to the need, BS 8903 gives guidance to the 'how'. It also gives guidance to the client-related opportunities for sustainability that lie through sustainable procurement whilst maintaining appropriate commercial interests (that difficult balance between commercial viability and, in some instances, sustainability of a project). The first part of the standard, the 'what', uses some familiar criteria that range through the three pillars of sustainability, to other more detailed criteria including climate change mitigation and adaption; water; waste; materials; biodiversity and ecology; land, air, water, and noise pollution; community benefits; transport and mobility; access; equality and diversity; health and well-being; ethics and labour standards; employment and skills; and SME (small- and medium-sized enterprise) support. The second part of the standard is centered on the 'why' or the business reasons for sustainable procurement. The standard rightly assumes that all relevant legislation is met and defines five main drivers for sustainable procurement:

1. **Financial**, including returns from reductions in waste and energy use.
2. **Marketing**: successfully marketing products in a global regulatory or reporting structure.
3. **Innovation**: An organisation that has the ability to produce workable innovative solutions to sustainable related problems is likely to thrive.
4. **Organisational values** and the standard drive companies and other organisations to ensure that their values are shared and enacted.
5. **Goodwill** of stakeholders, customers, investors, staff, clients, and shareholders (BSI 2010, reproduced by permission of BSI).

The third section of BS 8903 focuses on how sustainable procurement should be enacted, including centering upon the fundamentals, the enablers that allow the process to happen, and a more detailed look at the procurement process itself. The fundamental aspects include the organisational drivers which influence the organisational policies and strategies, which in turn feed into the procurement policy and strategy. This is likely to be very different depending upon the organisation; their relative importance needs to flow through the sustainable procurement process in the form of clear, and well-articulated, tailored objectives. Clarity is vital for staff to understand these objectives and act accordingly.

The enablers are the tools and techniques that allow a company to undertake robust, strategically appropriate procurement. This will normally include active and visible leadership that ensures that sustainability is embedded in governance processes across an organisation. One of the main challenges is that the leadership and governance also ensures that the commercial, technical, and business elements of the organisation's raison d'etre are fulfilled. This means getting the right people to champion this process and ensuring that they are competent in the field of sustainability, that is, having the right training, attitudes, behaviours, and tools for the job. Embedding these elements in staff appraisals and into personal objectives for the upcoming year is a method of linking the organisational objectives to the way in which staff act.

The standard also proposes similar links to external stakeholders where they can understand, through clear communication, the stance relating to sustainability from the client organisation and the expectations that are inherent through sustainable procurement. Through these stages the organisation will have maximised the likelihood of an effective sustainable procurement route and this is likely to lead to further, more effective, and wider engagement. The standard then establishes that the organisation needs to identify, prioritise, and manage risk for particular procurement routes and to measure and analyse the data that flow from the decisions made relating to sustainable procurement. This measurement element is important, for if certain variables are not measured the picture obtained doesn't fully describe the progress of the project's procurement. This in turn cannot enable good analysis and could contribute to poor future decisions. However, if the methodology is variable-heavy, too much data can sometimes be just as much a hindrance to trying to judge what is the truth behind the success (or not) of choosing a particular procurement route. The two main categories of measurement are (a) practice measures (departmental or company-based procurement practices) and (b) outcome-based measures (variables such as quantity and cost of energy saved, permanent jobs created, tonnes of waste diverted from landfill).

BS 8903 then uses a standard procurement process following these normal stages:

1. Identify the business need. This could be seen as the greatest opportunity for embedding sustainable procurement within an organisation. If a company or organisation decided not to procure a new building but decided instead to alter its internal practices or processes to meet its business needs, this might be one of the most sustainable decisions. This might not be good news economically for the construction industry, but this might be the best decision for the organisation in question. Therefore, it is important to understand in great detail the various possible alternatives and to challenge the possible 'obvious' decisions that might be normally made in situations such as this. Where a new building might be the normal method of procuring new space,

might leasing or sharing space be not only more cost effective but also more sustainable?

2. Define sourcing strategy. Looking for innovative solutions and defining a sourcing strategy that allows the organisation—within other appropriate boundaries such as cost, risk, time—to judge the efficacy of different sourcing routes, including life cycle issues. The specifications that flow from the sourcing strategy are important drivers to enable more sustainable buildings to be procured.

3. Identify suppliers and tender. BS 8903 suggests that the sustainability of the procurement is gradually built, starting with experience of the supplier/contractor related to sustainability in the prequalification questionnaire. This would then be followed by the request for specific competences related to sustainable practices, products, and outcomes within the tender documentation. This would then flow into the contract documentation. These elements would logically be included in the tender scoring specification, allowing sustainable procurement to be an embedded operation. Thus, many of the underlying principles of sustainable construction—low energy/high community and economic values; low waste/high performance buildings—that are discussed in the other chapters of this text are then used as methods of assessing whether to award a tender to a particular contractor. As part of this process, to help suppliers meet an organisation's needs, having a performance-based specification embedded in the aforementioned sourcing strategy will aid a flexible and possibly innovative response from suppliers.

4. Evaluate and award. At this stage of the procurement process the standard acknowledges that the supplier/contractor is particularly amenable to influence from the organisation procuring a building, and it is at this stage that the organisation can embed sustainable solutions, practices, and an end product. At this point, the standard recommends getting the contractor or subcontractor to understand the expectations of the organisation. As relationships tend to change once the contract is awarded, it is important that a good relationship built on reasonable expectations is forged at this stage.

5. Manage performance and relationship. BS 8903 suggests that sustainability indicators are monitored with the same fervor and with equal billing as the more normal business-related indicators, such as time, cost, and quality. In this way the performance monitoring process includes sustainability, rather than this being an adjunct to the main commercial operation.

6. Review and learn. The ability to continuously improve is the foundation of all good business practice, and this is also true when related to sustainable procurement. This field of endeavor is relatively new and as such provides many opportunities for the identification of new areas of best practice. Whilst buildings will always be very individualised products, lessons learned can give designers, contractors, and clients alike advantages when captured and used in their next project (BSI 2010, reproduced by permission of BSI).

2.3 BREEAM 2011 and sustainable procurement

The 2011 Building Research Establishments (BRE) Environmental Assessment Method, BREEAM 2011 (which replaced BREEAM 2008 on 1 July 2011), has a new section that addresses sustainable procurement. The section that alludes to sustainable procurement is MAN 01 and offers 8 new credits: project brief and design (4 credits), construction and handover (2 credits), and aftercare (2 credits). There are a number of guides that relate to this latest evolution of the BREEAM assessment series (further details in Chapter 5; assessment systems), and Table 2.3 is taken from the TRADA briefing of 2012 (TRADA 2012).

Once the process and the building have been assessed, the following credits are required (but not limited) to gain the range of ratings from Pass to Outstanding, shown in Table 2.4.

Whilst assessment systems have a particular focus—and this will obviously limit the extent the BREEAM 2011 (BREEAM International New Construction scheme was updated in 2013 and the UK-centric 2014 version saw a restructuring of the management section, with the minimum standard for issue Man 01 being removed and the minimum standards for issue Ene 01 adjusted) or other assessment systems can influence the extent of sustainable procurement—the

Table 2.3 A summary of the new items that relate to sustainable procurement within BREEAM 2011 since BREEAM 2008 (TRADA 2012)

BREEAM category	Description	Commentary
Sustainable procurement	1 credit for involving client, design team, occupier, and contractor in the design and decision making 3 credits for BREEAM AP being appointed and attending all key design team meetings 1 credit for a thermographic survey. This necessitates measuring water and energy use for 1 year, and holding a meeting with occupiers to explain, train, and hand over the building, complete with a helpline for the first 12 months.	8 credits plus 1 innovation credit. To obtain this credit the facilities management must make a commitment to provide BRE with 3 years of measured statistics on energy use, water use, and tenant satisfaction.

Table 2.4 Minimum standards for each rating to gain a pass for the sustainable procurement section of BREEAM 2011

BREEAM Rating	Pass	Good	Very Good	Excellent	Outstanding
Credit required	1	1	1	1	2

2.1 Want to know more? Procurement of Terminal 5, London Heathrow

Examples of representative procurement related to any building, either the procurement of the building as a whole, separate work packages, or for material and products, can be a bit of a minefield. By definition all buildings/sites are different: size, use, cost, complexity, client, designer, contractor, etc. However, there are some prominent projects that can be used as examples, as they show what can be done when working at the extremes of construction. If the largest, most costly, long-lived, and complex projects can be procured and managed through a collaborative system, then perhaps other less difficult projects can work just as well.

The Terminal 5 project at London Heathrow airport is such a project: large, complex, specialist in nature, a project that is of great interest to the specialist and mass media press who are liable to judge most elements of the finished building. Those aspects are likely to be related to economics, time, quality, fitness for purpose, and sustainability aspects. Terminal 5 was procured by BAA (British Airports Authority) with an aim of getting best value from the project and probably trying to satisfy the many previously stated aspects. From an analysis of BAA's previous project performance, two aspects that needed attention stood out: the lack of collaboration and the issue of responsibility for risk (Davies *et al.* 2009; Caldwell *et al.* 2009). Therefore, the obvious direction of procuring and managing the Terminal 5 project included addressing these two aspects and a partnering approach was adopted.

NOTE: The CEO of BAA at this time was Sir John Egan, who authored the influential UK construction report 'Rethinking Construction' http://www.constructingexcellence.org.uk/pdf/rethinking%20construction/rethinking_construction_report.pdf, a document that extols the same virtues of collaboration as adopted by BAA for Terminal 5.

The collaborative element of the procurement strategy allowed a number of the construction team to take the finance that they would have allocated to part of their tender and pool this resource to pay for the risk-associated costs. This allowed the smaller subcontractors who normally ended up bearing much of that risk to be competitive but also have some certainty against some of the more risky aspects of the project (Doherty 2008). Within T5, BAA sought to create a new culture that encouraged people to

- Seek out, capture, and exploit the best practices of others
- Remove the barriers and inhibitors to doing things differently
- Stimulate and support good ideas
- Leverage the commercial incentive to perform exceptionally (Wolstenholme *et al.* 2008)

Partly through the procurement route and the changes in ethos of the project, T5 allowed the setting and delivery of new standards in environmental sustainability for the construction industry at the time of construction. Opportunities at each development stage were taken to embed environmental awareness and corporate responsibility into decision-making processes. Throughout the design and construction, project teams and suppliers were encouraged to apply innovative techniques and best practice to deliver exemplary environmental performance (Lister 2008).

there is a very accessible account in Womack and Jones (2003). Value as a concept is discussed, as well as value flows that should be focused on the end product. The definition of value associated with a building or project is key to value management, a term that describes a process that uses meetings/workshops, interviews, and reviews to evaluate the requirements of a project against the possible methods that could be used to fulfill the needs of the project. According to the UK's Constructing Excellence, value management should 'be an integral part of any projects viewed on a whole life basis' (Constructing Excellence 2004), and the benefits of value management and lean management within the construction industry have been widely discussed (see Ōno 1988, Vrijhoef & Koskela 2000, Rother & Shook 2003, Constructing Excellence 2004, Womack et al. 2007, Kelly et al. 2008, Designing Buildings 2013).

2.4.2 The project brief

The project brief can be one of the most important documents within the procurement process. It is the connection between the client's needs/requirements and the production of the building/project in a sustainable fashion. Much as the careful placing of the good foundations for a building can influence the success of the next stages of the building process, a well thought through brief can communicate the vision, ethos, and direction of the project as a whole. The project brief also needs to clearly state where the sustainability of the project lies, where this will be measured, and how the outputs relate to the client's objectives. Much as all clients and sites are different, so the objectives in individual project briefs differ. It is rare that a previous brief in its entirety can be used a second time. The client's requirements need to be discussed at the earliest stage of the inception period with the appropriate construction professional (e.g., architect or project manager) to separate the essential from the desirable. These can then be mapped against the usual design objectives of any project, including where appropriate, the business case in its fullest sense (normally part of an earlier internal client related process), any policies that flow from the client or parent companies, social/community related objectives, specific whole-life issues, and locational/transport issues.

2.4.3   Producing tender documents including output-based specification

Much of the format and content of the tender documentation will be informed by the decided procurement route project brief and defined outputs that will ensure the project is sustainable.

Some of the issues that can dramatically influence the sustainability of the finished building can rest upon who is responsible for some of the specified outputs. For example, a service element, such as energy consumption of the finally completed building, can rest with various parties: the client, if an owner occupier; the tenant (if the building is rented); and sometimes the supplier of the building, if a PFI (Private Finance Initiative, a UK government term related initiative) or design, build, and operate agreement. More innovative agreements include those where the parties related to the owning, design, building, and operation can share the 'pain' or the 'gain' of making or underachieving efficiency specifications. In this way many more parties feel they have a direct relationship with the goals of the client/occupier and will encourage their employees and representatives to make good decisions and cope more readily with poor decisions made by other parties by sharing the burden, financial or otherwise.

Parties that have been invited to tender should be asked to provide full details of how they will respond to the required sustainability objectives in a wider sense. The importance of this element in the tender appraisal process should be made clear to reinforce what will be expected if they are successful. The following should be done:

- The construction team's knowledge and experience of sustainable projects is evaluated.
- The construction team's suitability is assessed against sustainability criteria in a clear and methodical manner.
- The construction team is encouraged to suggest innovative approaches and alternatives that offer better value for money and/or whole-life cost performance on behalf of the client. This might include offering instances of where a team member has improved the performance of past projects.

2.4.4   The integrated supply team

A supply chain for any construction project that does not buy in to the procurement goals of the client, design, construction, and maintenance teams is very unlikely to result in 'sustainable procurement'. As described previously, the ability to influence the behaviour of the companies that make up part of the wider team of a larger construction project is a vital approach that influences the supply chain. It can help establish clear expectations related to purchasing choices linking the actions of purchasing products and services and maximising the sustainability of the whole. Without the communication of each organisation's environmental policies and practices to each other at appropriate stages of a project, these expectations will be vague. However, these documents can be quite generic, often written and then provided when requested, rather than adhered to. Actions are needed to allow the content of these to be common knowledge and ramifications on site made clear. Output specifications, perhaps based on metrics such as energy consumption per metre squared, the maintenance of an interior temperature for given exterior conditions, or a total of embodied energy in certain material categories are one method of establishing the performance of a building. However, if the project team does not fully buy in to these at an early stage, suppliers are likely to be uninformed of assumed expectations, reducing the likelihood of meeting the specification. Some methods of ensuring the systems used within contributing organisations include requirements for use of environmental management systems, such as 14001 and other quality assured standards (many of which are described in Chapter 6).

2.5   Contracts and sustainable construction

Agreements between parties, if genuine and communicated in appropriate ways, can form a very good basis for ensuring that all those elements that are of lesser financial importance but vital to achieving a sustainable constructed product are addressed. Unfortunately, many contracts are seen as a basis of settling disputes and not as a mechanism to clarify roles, responsibilities, and penalties (that would be imposed if unsustainable actions were taken).

The UK's leading supplier of construction contracts, the Joint Contracts Tribunal (JCT), has produced *Building a Sustainable Future Together*, a guidance note that is intended to assist clients, management, and design teams, to place an environmental/sustainability ethos within contracts used in the procurement of construction projects (JCT 2011). This document focuses, in particular, on how sustainability requirements can be provided for in the contract documentation, as well as the following key areas:

- Procurement
- Contractual approaches to sustainability
- JCT's sustainability provisions
- Framing the detailed requirements
- Building use and maintenance
- Evaluation

Further support of early involvement of the construction team is provided by JCT's assertion, 'The client's commitment and the early involvement of the supply chain are necessary to achieve sustainability both in the design and construction processes' (JCT 2011).

This requirement of the early involvement of parties is supplemented by a series of other sustainability-related clauses. For instance, in addition to requiring the contractor not to use or approve certain products (products and materials identified as being potentially hazardous or not conforming to various publications from the British Council for Offices, British Property Federation, relevant British or European standards or codes of practice, or the Building Research Establishment), there is also a contractual term requiring the contractor to ensure the building achieves an excellent BREEAM rating. This places responsibilities, for the sustainability of any project using this contract with the guidelines, upon the contractor. A series of case studies showing how the contractual arrangements have impacted upon the sustainability and general success of a range of buildings can be found on the JCT website (JCT 2013a).

Additionally, JCT recently consulted with various elements of the construction industry concerning the use of more explicit sustainability terms within its own contract forms. This resulted in the JCT report *Sustainability: Lifecycle Consultation* (JCT 2013b). The results were very interesting and reflect a mix of those respondents who wish the contractual forms to remain much as they are and those wishing to challenge the system a little more.

As discussed, the early appointment of contractors and other directly associated construction professionals are an ongoing upgrade in the process of procuring buildings and civil engineering projects. Many of the more recent contractual arrangements, particularly those that are based upon partnerships, such as PPC2000 (Project Partnering Contract) and the associated SPC2000 (Specialist Partnering Contract) and TPC2005 (Term Partnering Contract) facilitate this upgrade. It can be noted that many of these standard forms of contractual agreement flowed from Sir John Egan's 'Egan Report', including the use of preconstruction phase agreements (Egan 1998).

NEC, the New Engineering Contract, was launched in 1993 and was praised in the 'Latham Report' (Latham 1994). An updated set of documents (NEC2) was released in 1995, and the current (but being updated by addition) series, NEC3, was introduced in 2005 (NEC 2007). NEC3 is a family of contracts that embeds many project management practices while defining the legal relationships and responsibilities between parties. NEC3 promotes a partnering, collaborative approach that echoes parts of the guidance from BS 8903 alongside the aspects of client satisfaction such as delivering projects to budget and on time (NEC3 2013).

2.6   The RIBA plan of work

The Royal Institute of British Architects (RIBA) 'Plan of Work' was introduced in 1963 and is the UK model for defining the stages of building design and the construction process. Using these defined stages, the Plan of Work is used by the predesign, design, production, and postoccupancy stages of a building project. Through these stages the Plan of Work can be the basis for a process map and a management tool, reflecting the many reference points used by many industry-standard documents.

The latest version of the RIBA Plan of Work, the 2013 version, has a number of fundamental changes that are likely to allow it to be a contributor to more sustainable buildings (RIBA 2013). According to RIBA's webpage dedicated to the new Plan of Work, there are a series of beneficial aspects:

- Eight key work stages instead of 11
- Suitable for all sizes of projects and practices
- Suitable for all types of procurement routes
- Available online to enable customised versions
- Clear, comprehensible, and flexible for users
- Reflects the complexities of developing and delivering construction projects in the 21st century

The online aspects of the new toolbox reflect the move towards Building Information Modelling and the driver that has become ubiquitous throughout this section, that of the need for collaboration. The toolbox comprises a customisable series of spreadsheets, matrices, and schedules.

2.6.1 Green overlay for 2007 RIBA plan of work

A document that reflects sustainability's place in the overall sequence of events (predesign, design, production, and aftercare of a building) is the 'Green Overlay to the RIBA Outline Plan of Work', for the previous version of the RIBA Plan of Work (Gething 2011). The overlay inserts a series of words or pieces of text that 'illustrate behaviours and activities that will support a more sustainable approach' (Gething 2011). They range from smaller insertions, such as more sustainable links to the Plan of Work placed in the text, to 'Sustainability Checkpoints'. The fusing of sustainability-driven goals within the context of a well-used and respected document can be a powerful driver for change. Therefore, an addition to an original document is probably the best method of underlying the value of sustainable thinking throughout the procurement process to the use of the building.

2.7 The sustainable procurement of materials and equipment

Whilst the overall framework of sustainably procuring a building is finding a more confirmed structure, built environment professionals are still left with the task of isolating the 'right' products, materials, and services that provide an optimal solution, including the sustainability credentials on the ground. For larger companies the leeway in stepping outside commonly used purchasing routes can be small. As previously discussed, partnership agreements, company policies, and nominated suppliers can all 'lock in' choice. This can be further limited through client requirements, budget, design strategy, and so on.

Whilst not an organisation that is solely focused upon construction, the Sustainable Purchasing Leadership Council, officially launched in July 2013, promises to aid the more detailed elements of sustainable procurement. Its stated aim is 'of accelerating the market's adoption of the sustainable purchasing best practices' (Sustainable Purchasing Leadership Council 2013). The council has a number of influential partners including the American National Standards Institute (ANSI), the Association for the Advancement of Sustainability in Higher Education (AASHE), Business and Institutional Furniture Manufacturers Association (BIFMA), ICLEI (a non-profit membership organisation that has been working on sustainable procurement for 17 years), Institute for Supply Management (ISM), ISEAL Alliance (the global membership association for sustainability standards), the National Association of State Procurement Officers (NASPO), Practice Green Health (PGH, a US leading health care community), the Product Stewardship Institute (PSI), and the Sustainable Food Lab. Whilst being a relatively young organisation, this council promises the production of tools that could be used as part of a sustainable procurement strategy for construction projects.

2.7.1   Materials and product-related guides

There are a number of guides and processes that can be used to isolate a number of different materials and products that would be seen as suitable alternatives for use in a sustainable building. They might warrant their inclusion through good environmental credentials (in use and through production and disposal), proven durability, lower cost, ethical production, and an overall rating through life cycle analysis. These guides can help the reader to drill down to a more detailed series of decisions below an umbrella of overall sustainable procurement. The details and theories behind these systems and procedures contained in these guides can be found in the materials section of Chapter 4.

The UK's BRE developed its first edition of the 'Green Guide' series in 1996. This was aimed to provide a guide to the relative environmental impacts of different building materials, based on numerical data. This guide has been constantly developed over many years and now is available online (BRE 2013). The guide's environmental rankings are based on life cycle assessments, which is, in turn, based on the BRE's Environmental Profiles Methodology 2008. Specific materials and components may be located using the building element that they are ostensibly part of: roof, floors, walls, and so on. Final 'Green Guide' ratings come from a combination of the materials performance related to a series of categories, such as water extraction, climate change, waste disposal, and a further 10 categories. This data is ranked from A+ to E, where logically A+ represents the best environmental performance/least environmental impact, and E the worst environmental performance/most environmental impact.

A well-respected text that has helped building professionals select and procure appropriately green materials over a number of years is the *Green Building Handbook: A Guide to Building Products and Their Impact on the Environment* (Woolley & Kimmins 2000). This text, whilst now a little dated, describes a good selection of materials alongside their environmental credentials and an articulation of many of the ideals behind green architecture.

Recently (as described in the materials section of Chapter 4) the UK's BRE developed BES 6001 (BRE 2009), a responsible sourcing standard for construction products. The standard describes a framework for governance, managing a supply chain, and the environmental and social aspects that must be addressed in order to ensure the responsible sourcing of construction products.

2.8   Summary

Step one in the process of obtaining a sustainable building is to ensure that the necessary elements of sustainable procurement have been followed. The client and his or her advisors need to establish whether constructing a new or refurbishing an existing building is in that organisation's best interest. Following this, a procurement route and then an appropriate brief need to be established, building in all those vital elements that might deem any project unsustainable. The use of an integrated construction and supply team, using appropriate contractual arrangements, should get the project off to a good start. However, the building needs to be designed, constructed, and managed in appropriate ways to ensure that all the good intentions built into the procurement stage are not wasted. In the next chapters these aspects will be discussed.

References

Action Sustainability (2012a) *Sustainable supply chain diagnostic* [Online]. Available: http://www.actionsustainability.com/evaluation/sustainable-supply-chain-diagnostic/ [16 Dec 2013].

Action Sustainability (2012b) *Supply chain sustainability* [Online]. Available: http://www.supplychainschool.co.uk/about/about-our-partners/skanska.aspx [27 Mar 2014].

BRE (2009) *BES 6001: ISSUE 2.0 Framework Standard for the Responsible Sourcing of Construction Products*, BRE Global, Watford.

BRE (2011) *BREEAM 2011, 4.0 Management, Man 01 Sustainable Procurement* [Online]. Available: http://www.breeam.org/BREEAM2011SchemeDocument/Content/04_management/man01.htm [17 Dec 2013].

BRE (2013) *Green guide to specification* [Online]. Available: http://www.bre.co.uk/greenguide/podpage.jsp?id=2126 [17 Dec 2013].

BSI (British Standards Institute) (2010) *BS 8903:2010 Principles and framework for procuring sustainably: Guide* BSI, London.

CABE (Commission for Architecture and the Built Environment) (2009) *Agreeing on a procurement strategy*, CABE, London.

Caldwell, N. D., Roehrich, J. K., & Davies, A. C. (2009) Procuring complex performance in construction: London Heathrow Terminal 5 and a private finance initiative hospital, *Journal of Purchasing & Supply Management* 15, 1478–4092.

Constructing Excellence (2004) *Value constructing excellence* [Online]. Available: http://www.constructingexcellence.org.uk/pdf/fact_sheet/value.pdf [17 Dec 2013].

Davies, A., Gann, D., & Douglas, T. (2009) Innovation in megaprojects: Systems integration at London Heathrow Terminal 5, *California Management Review* 51, 101, ISSN:0008-1256.

Designing Buildings (2013) *Case study: Terminal 5 Heathrow* [Online], Available: http://www.designingbuildings.co.uk/wiki/Procurement_of_Heathrow_T5#Procurement_Systems [17 Dec 2013].

Doherty, S. (2008) *Heathrow's Terminal 5: History in the Making*, Wiley, Chichester, UK.

Egan, J. (1998) *Rethinking construction: Report of the Construction Task Force to the Deputy Prime Minister, John Prescott, on the scope for improving the quality and efficiency of UK construction.* DTI, HMSO, London.

Emmerson, H. (1962) *Survey of problems before the construction industries*, Ministry of Works HMSO, London.

GCCP (Government Construction Clients' Panel, Sustainability Action Group) (2000) *Achieving sustainability in construction procurement* [Online]. Available: http://webarchive.nationalarchives.gov.uk/20110601212617/http://www.ogc.gov.uk/documents/AchievingSustainabilityConstructionProcurement.pdf [31 Aug 2015].

Gething, B. (ed) (2007) *Green overlay to the RIBA Plan of Work*, RIBA, London [Online]. Available: http://www.architecture.com/Files/RIBAProfessionalServices/Practice/General/GreenOverlaytotheRIBAOutlinePlanofWork2007.pdf [27 Mar 2014].

HMS Government and Strategic Forum for Construction (2008) *Strategy for sustainable construction* [Online]. Department for Business, Enterprise and Regulatory Reform, London. Available: http://www.bis.gov.uk/files/file46535.pdf [16 Dec 2013].

Hughes, W., Hillebrandt, P. M., Greenwood, D., & Kwawu, W. (2006) *Procurement in the construction industry: The impact and cost of alternative market and supply processes (Spon Research)*, Taylor & Francis, New York.

JCT (Joint Contracts Tribunal) (2011) *Guidance note: Building a sustainable future together*, JCT, London.

JCT (Joint Contracts Tribunal) (2013a) *Case studies* [Online]. Available: http://www.jctltd.co.uk/case-studies.aspx [17 Dec 2013].

JCT (Joint Contracts Tribunal) (2013b) *Sustainability: Lifecycle consultation report 2013* [Online]. Available: http://www.jctltd.co.uk/docs/JCTSustainabilityLifecycleConsultationReport2013.pdf [17 Dec 2013].

JCT (Joint Contracts Tribunal) (2013c) *Welcome: Procurement* [Online]. Available: http://www.jctltd.co.uk/procurement.aspx [17 Dec 2013].

Kelly, J., Male, S., & Graham, D. (2008) *Value management of construction projects*, Wiley-Blackwell, Oxford.

Latham, M. (1994) *Constructing the team: Final report of the government/industry review of procurement and contractual arrangements in the UK construction industry*, HMSO, London.

Lister, B. (2008) Heathrow Terminal 5: Enhancing environmental sustainability, *Proceedings of the ICE: Civil Engineering* **161**, 21–24, [Online], Available: http://www.icevirtuallibrary.com/content/article/10.1680/cien.2007.161.5.21 [10 Dec 2015].

Morledge, R., & Smith, A. (2013) *Building procurement*, Wiley-Blackwell, Oxford.

National Audit Office (2005) *Improving public services through better construction.* Available: https://www.nao.org.uk/report/improving-public-services-through-better-construction/.

NEC (2007) *NECP/B01* [Online]. Available: http://www.neccontract.com/documents/panel%20briefings/NECPB01%20-%20Culture%20&%20Mindset%20webcopy.pdf [16th March 2014].

NEC3 (2013) NEC3 contract suite, NEC Engineering and Construction Contract website, [Online], Available: https://www.neccontract.com/ [6 Dec 2015].

OGC (Office of Government Commerce) (2007) *Procurement and contract strategies: Achieving excellence in construction procurement guide* [Online], Available: http://webarchive.nationalarchives.gov.uk/20110601212617/http:/www.ogc.gov.uk/ppm_documents_construction.asp [25th May 2015].

Ōno, T. (1988) *Toyota production system: Beyond large-scale production*, Productivity Press, New York.

RIBA (Royal Institute of British Archiectects) (2013) *Plan of work* [Online], Available: http://www.ribaplanofwork.com/Toolbox.aspx?utm_source=bookshops&utm_medium=referral&utm_campaign=PoW_2013 [17 Dec 2013].

Rother, M., & Shook, J. (2003) *Learning to see: Value-stream mapping to create value and eliminate Muda, Version 1.3*, Lean Enterprise Institute, Cambridge, MA.

Sims, N. (2006) Foreword. In: *Procuring the future, sustainable procurement national action plan: Recommendations from the Sustainable Procurement Task Force*. Department for Environment, Food and Rural Affairs, London.

Skanska (2011) *Skanska sustainable procurement* [Online]. Available: http://www.skanska.co.uk/Global/About%20Skanska/supply%20chain/Skanska%20Sustainable%20Procurement%20Policy%20May%202011.pdf [17th March 2014].

Strategic Forum (2012) Construction commitments [Online]. Available: http://www.strategicforum.org.uk/cc.shtml [16 Dec 2013].

Sustainable Purchasing Leadership Council (2013) History of the council [Online]. Available: http://www.purchasingcouncil.org [17 Dec 2013].

TRADA (2012) *Construction briefing BREEAM: A summary of changes to the 2011 version*, TRADA, High Wycombe.

Vrijhoef, R., & Koskela, L. (2000) The four roles of supply chain management in construction, *European Journal of Purchasing & Supply Management* **6**(3), 169–178.

Wolstenholme, A., Fugeman, I., & Hammond, F. (2008) Heathrow Terminal 5: Delivery strategy, *Proceedings of the ICE: Civil Engineering* **161**(5).

Womack, J. P., & Jones, D. T. (2003) *Banish waste and create wealth in your corporation*, Free Press, New York.

Womack, J. P., Jones, D. T., & Roos, D. (2007) *The machine that changed the world: The story of lean production—Toyota's secret weapon in the global car wars that is now revolutionizing world industry*, Simon and Schuster, London.

Woolley, T., & Kimmins, S. (2000) *Green building handbook: A guide to building products and their impact on the environment, Vol. **2***, Spon Press, London.

3. Energy, water, refurbishment, and sustainable building design

3.1 Design-related sustainability

As described in Chapter 1, the concepts underpinning sustainability are relatively straightforward; difficulties arise when these concepts are applied to real situations. Things don't get much more real than decisions relating to actual buildings on real sites with real people occupying them. This chapter describes three of the major design-related elements that impact the sustainability of such a building: energy, water usage, and refurbishment. The specification and use of materials in building design is described in Chapter 4. Other issues, such as design philosophies, wider planning issues, and global emission pressures, are touched upon where necessary but are not the major focus of this chapter.

3.1.1 Design and construction: the two main areas of interest

Before linking the process of design with the sustainability of buildings and structures, it is helpful to know what design actually means in this context. Dictionaries are a good place to start, but they offer broad definitions such as 'a mental plan' or 'the art or action of conceiving of and producing a plan or drawing of something before it is made' (Oxford Dictionaries 2014), neither particularly helpful for the use of this text. 'Design' in the context of buildings used in this book refers to the processes and stages of processes that will precede the production of a building or structure, but not the process of construction nor the aftercare and deconstruction. That design decision-making process, although specifically focused on the time before construction takes place (in this text's definition), will have impacts that endure. This is because the creation of a building, or structure, uses resources such as land, materials, and people's

Sustainable Construction Processes: A Resource Text, First Edition. Steve Goodhew.
© 2016 John Wiley & Sons, Ltd. Published 2016 by John Wiley & Sons, Ltd.

time and capital, all of which could be spent on other things. A building provides an internal environment using resources over time and then dealing with the waste, almost as if the building were a living organism. The envelope of any building provides shelter, acting as an environmental filter, enabling some control over heat, light, noise, moisture, air, and air-borne pollutants.

Design, therefore (unsurprisingly), directly influences the environmental, economic, and social performance of the building, both in its immediate surroundings and further afield through emissions with global impact. Buildings will have to continue to function under a changing climate, requiring them to respond to heating requirements, cooling and ventilation, and to be robust to flooding or material damage. For these reasons, the many separate aspects of construction design will be investigated in this chapter. Other impacts that derive from materials and the construction process will be dealt with in Chapters 4 and 5, respectively.

3.2   Sustainable design

There are many facts, figures, and statistics that demonstrate how the design of buildings and structures influences the sustainability of our built environment. Some of the more general highlights are listed in Table 3.1.

In summary, the construction industry is big, its products are indispensable, and the design of those products directly affects people's lives.

Some decisions that relate to the sustainability of a construction project are out of the designer's hands. The project site might not be ideal, either for making good use of natural phenomena for heating/cooling/ventilation, or because it is sited far from a potential workforce or workplace. Clients and planners may have requirements that conflict with some principles of sustainable design.

Table 3.1 Drivers for sustainability in construction

Driver	Impact
Size of the construction industry	The construction industry in the UK accounts for approximately 100 billion GBP per annum or 8% GDP (UK Department for Business and Skills 2008) http://www.berr.gov.uk/whatwedo/sectors/construction/sustainability/page13691.html). An industry of this size can have a large impact upon the sustainability of the UK as a whole.
Product	As buildings and infrastructure are a necessity for any modern society, products of the UK construction industry and their design are important for achieving sustainable activity in other economic sectors.
Behaviour	If the buildings in which people live, work, and shop are designed to be sustainable, it is likely that people will find it easier to live a more sustainable lifestyle in other ways—in the ways they travel, deal with waste, and even spend their leisure time.

Table 3.2 Aspects of construction associated with a sustainable design focus

Sustainability Aspect (in no particular order)	Design Factors
Environment	Minimise operationally related emissions. Maximise internal comfort. Minimise embodied energy contained in materials, energy in construction, and energy expended through maintenance procedures. Design to allow for the building to adapt to a changing climate.
Economic	Design for use of local materials. Design for low maintenance. Design to reduce the running costs of the building or structure. Compare the differential cost of dwelling units with the cost of the surrounding units. Consider the nature of the tenure, degrees of shared ownership, and allowance for key workers.
Social	Ensure comfortable and healthy interior environments. Design for ease of construction to minimize accidents and injuries. Strive to design a building or structure that delights.

What is sustainable design when related to sustainable construction? It is more than producing (through good design) buildings and structures that have minimal impact upon the global, local, and internal environments. Sustainable design should also have regard for the economic and social impacts upon the communities that will build, use, and maintain that building or structure (Table 3.2). Contemporary issues of climate adaptation and mitigating the effects of climate change need to be taken into account. The design process should also ensure that the demolition or dismantling of the building is straightforward, minimising environmental damage and the energy used to deconstruct the building.

As discussed in Chapter 2, one must ask the question, does a need (commercial, social, economic, personal) justify constructing anything in the first place? Once this rather tricky question has been debated and the answer is in the affirmative, then the broader issues of where and what should be constructed can be tackled.

3.2.1   Location of the construction development

The site location has an impact on the sustainability of any construction project. This impact ranges from transportation issues to access and natural energy sources. Table 3.3 shows some of these site issues.

Whilst accepting that land is a finite resource and therefore locations are never perfect, the choice, if one exists, of one spot or another will strongly influence the performance of the building(s) long before any parameters relating to fabric, energy, shape, size, form, or use are considered by a designer.

Table 3.3 Location issues related to construction development

Locational Issue	Design Considerations for Sustainability
Transportation	Social (access to services and amenities) Economic (access to workplace and to markets) Environmental (emissions from travel, new transport routes required, and knock-on construction issues) Logistics, or will the site be accessible to potential flooding?
Natural resources	Orientate/configure a building's shape and size to take full advantage of the sun, wind, water, and other building's/topographical features. Location of building materials. Ensuring the natural capital (for more information relating to the definition of *natural capital* please see Chapter 5) of a site is at the very least maintained or preferably enhanced
Commercial advantage	Access to commercial partners and potential markets, along with lack of competition can impact upon the long-term success of a project,
Planning issues	Zoning of the area, the acceptance of the building or structure where it sits in the urban, semi-urban, or rural landscape Historical influences, listings? Conservation area? What can the design team incorporate into the design on the development that is beyond the minimum required by planning guidance or legislation? Density of the build form; how many units can comfortably sit within the site's boundaries whilst still keeping the individual and the community happy and coherent?

3.3 Energy and design: Building fabric

3.3.1 Energy in use

Our use of energy has, and will continue to be, central to our ability to maintain our living standards, working patterns, and leisure time, but energy and its use is linked to environmental damage, social influence, and economic consequences. Buildings consume energy through their use and contain energy that is embodied in the construction and materials that make up the building's fabric and services.

Buildings directly or indirectly account for approximately 45% of all the energy used in the UK, so this issue is a major one for sustainable construction. This section examines the aspects of energy use and production that mesh with the sustainable design process (energy and the process of sustainable construction are reviewed in Chapter 5).

Energy is surprisingly difficult to define for those not of a scientific persuasion. We come across energy as a secondary issue when the need to travel, to illuminate the house, or to cook a meal arises. Classically, energy is defined as something that has the capacity to undertake work and is measured in the SI unit: the Joule. The Joule can be used to measure or describe many different forms of energy: electrical, heat, mechanical, and chemical, and it is often the conversion between these energy forms that is of interest. Kinetic energy (energy associated with a mass in motion) in the form of a moving automobile can generate heat in its brakes when slowed.

Potential energy (energy associated with a mass held a distance above the earth's surface) in the form of an elevator counterweight held at high level, for example, can provide kinetic energy in the form of an elevator car moving upwards. The rate of energy use (power) is measured in Joules per second or Watts. By measuring the power used over a period of time, the total amount of energy consumed can be calculated and compared (normally in kWh). In this way different forms of energy use can be combined and the final totals compared with another building or process.

3.3.1.1 Energy conversion

To maintain a building at a comfortable internal environment, it is obvious that in the colder months of the year heat is required, whereas in the summer months cooling may be necessary. There are various methods to obtain the energy from different sources: chemical conversion (combustion), an electrical source (passing an electrical current through a resistor), or even a mechanical source (through the use of friction). The decision to use a particular primary source of energy will be influenced by many different factors, such as the availability of the primary source and the efficiency/by-products of the process used to convert that primary energy into the energy form required. A typical example is the use of fossil fuels to generate electricity for use in homes, industry, and transportation. Let us assume that quantities of a fossil fuel, either coal, oil, natural gas, or another derivative, are to be burnt (chemical energy) to raise steam, which in turn will be used to operate a steam turbine (mechanical energy), which in turn will produce electricity (electrical energy) by rotating an electrical generator. This electrical energy can then be used in buildings by converting it to light, heat, and mechanical energy. At each point of energy conversion and in energy transportation, a percentage of the primary energy will not be transferred and will be effectively lost. Figure 3.1 shows the stages of energy conversion and transportation.

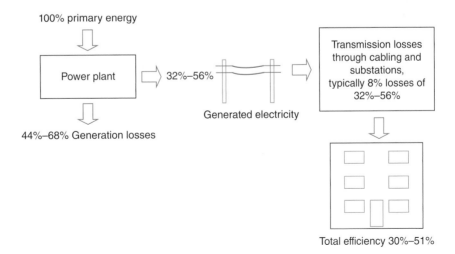

Figure 3.1 The energy conversion percentages for electrical generation and transportation using fossil fuel–based primary energy (US Department of the Interior 2014).

Table 3.4 Primary energy demand, 2012 (Department of Energy and Climate Change (2015a)

Energy Use/Loss	Percentage of Total Electrical Energy Generated
Final energy consumption	67%
Non-energy use	4%
Use by energy industries	6%
Losses in transformation	21½%

Recent figures for the UK's final energy consumption related to the losses in electrical consumption came to approximately 35% (as can be seen in Table 3.4) when measured from the point at which the electrical power enters the distribution system and enters the building.

The inevitable conclusion here is that if energy is generated through many stages of energy conversion and transported over great distances, it will result in a series of inefficiencies that will greatly reduce the amount of primary energy that actually gets to a building. This fact needs to be taken into consideration when decisions are being made that affect the energy use in a building or the energy used during the construction process. If you switch on an appliance in a building, it consumes a quantity of energy that represents only a percentage of the primary energy used to generate that electricity. This is not only an inefficient use of a scarce resource but also an effective multiplier of the amount of greenhouse emissions from fossil fuel (assuming fossil fuels were used to generate the electrical power). For these reasons, using a finite resource through a process that is inherently inefficient cannot be the most sustainable way to supply the energy needs of buildings. The next logical question is, how else can we do this? There are a large number of alternative methods of generating and supplying energy. Some are radical, whilst others are less so but often no less controversial. Some may be seen as temporary stepping stones from our fossil-dependent present to a more efficient and clean future energy solution.

None of the current energy generation and supply scenarios are being advocated by the majority of global authorities as a sole solution. A 'mixed economy' of different solutions is most likely, as each has its advantages and disadvantages. This implies that the efficiency and therefore the associated emissions from the generation of electrical energy is likely to improve—but slowly. Buildings and structures will have their part to play, either as a platform for stand-alone generation or as one of the largest consumers of energy. As the energy infrastructure evolves, so too will the design of a sustainable building.

3.3.2   Overall energy consumption

The energy consumption of the different broad sectors of the UK can be seen in Tables 3.5 and 3.6.

Table 3.5 Final energy consumption by use, 2014 (DECC 2015b)

Industry or Sector	Percentage of Total UK Energy Consumption
Transport sector	39%
Domestic sector	30%
Industry	17%
Other users	14%

Table 3.6 The energy consumption of the different broad sectors of the UK (DECC 2015a)

Fuel/Energy Types Used in the UK, 2012	Percentage of Total UK Energy Consumption
Petroleum	47½%
Electricity	18½%
Natural gas	28%
Other*	6%
Total	142.8 million tonnes of oil equivalent

*Includes coal, manufactured fuels, renewables and waste, and heat sold.

As buildings consume such a large fraction of the energy generated in the UK, it is wise to scrutinise the facts behind this consumption. Of the building stock, which buildings consume more than others? Is the quantity of energy usage dependent upon building type and size? What of factors such as the age of buildings, geographical location, occupancy patterns? In reality, energy consumption is related to a combination of these factors. One prominent truth strikes home: no matter what the cause of high or low energy consumption, if buildings can be designed to consume less energy, the impacts of energy generation will be reduced. For this reason, it is important to look into how sustainable design and construction can reduce this consumption. The breakdown of where energy is used in buildings can be seen in Figure 3.2.

As can be seen in Figure 3.2, the major uses of energy in UK commercial buildings are space and water heating, comfort cooling, and lighting. As UK domestic buildings tend not to have comfort cooling and tend to be more effectively lit naturally, their main energy consumption is for space and water heating, as seen in Figure 3.3.

The strategies for reducing energy consumption in both sectors are similar. A connected series of factors will influence the energy consumption of any particular building.

Many of the factors that reduce the energy consumption of a building, listed in Table 3.7, can be taken into account during the design phase for a proposed new building. The extent of these and the way in which they are integrated into the building's design hinge on the philosophy of the design solution and the requirements of the client/occupant. Low-energy buildings often perform best when an integrated approach is applied ensuring the building fabric is designed to fundamentally reduce the need to heat,

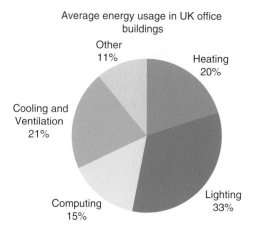

Figure 3.2 The breakdown of where energy is used in buildings (http://www.cibsejournal.com/cpd/2010-11/).

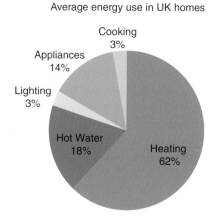

Figure 3.3 Average energy use in UK homes (Palmer & Cooper 2012).

light, ventilate, and otherwise service interior spaces. Once the generic aspects described in Table 3.7 have been taken into account, it is logical to turn to the building fabric and its impact upon the building's energy performance. Since heating and cooling are currently the largest consumers of energy and are influenced by any building's fabric, the next section of this chapter is devoted to this aspect of building performance.

3.3.3 Sustainable heating and cooling

Buildings are built and used for many different purposes. The requirements for shelter and comfort are supplemented by needs or desires for work space, production facilities, security, storage, investment, and in some instances 'just

Table 3.7 Generic factors affecting energy consumption of buildings

Factor Affecting Energy Consumption	Influence of the Factor
Building design	Large variances in energy consumption can result from different design strategies. Strategies vary from different thermal insulation values to the ability of a building's façade to make use of heat from solar gains and illumination from daylight.
Age	The use of improved technologies and materials in insulation and technologies associated with building services are likely to mean that older buildings will consume more energy than new ones.
Legislation	Over time new legislation that governs the thermal characteristics of the fabric and services of a building have improved buildings' efficiency. A building built just before energy-related legislation (such as Part L Approved Documents of the Building Regulations) is likely to consume more energy than a building built after the enactment of the legislation.
Size	Although an obvious point, a larger building is likely to consume more energy than a smaller building, even very efficient large buildings.
Occupation patterns	If a building has methods of turning off or down the energy needed to function when spaces are unoccupied, then energy savings can be achieved.
Occupant behaviour	Two identical buildings occupied over the same time period for the same use can still have substantial differences in their energy consumption.
Geographical location	A building built in a part of the UK where the number of degree days (the number of days when heating is required to maintain a comfortable normal internal temperature) are considerably higher than another will normally use more energy to heat the building.

having a base' in a particular location. However, there is one truth: occupants or machinery need to have a thermal environment that will allow them to live/function to their best potential. Historically, buildings needed to rely upon natural processes to maintain their internal conditions. Heating was often limited to open hearths, one or more in each room, sometimes linked with cooking. This was the position for commercial buildings until the introduction of central boiler systems into hotels and higher-class office buildings, where a central heat source supplied hot water for heated rooms via separate heat emitters placed strategically around the building. This type of central heating system migrated to UK domestic buildings, mostly over the late 1960s and early 1970s. Today's office spaces and shopping premises use systems to cool our buildings, often in the form of air-conditioning, as the norm rather than the exception. However, this is at the cost of energy and emissions; it can be expensive, and it adds complexity and sometimes build-time. Many modern buildings 'do what they are told' if the building management and control systems are functioning correctly. But there are new ways of using traditional heating and cooling systems along with modern technologies to attain low energy use and high levels of occupant comfort.

Exact values that describe the boundaries of an environment that matches human requirements can verge on the 'one size fits all' school of design. What temperatures, humidities, air speeds, or air quality values should be the goal of the building design team? One could also consider the human interaction with the internal environment of a building. From human comfort pioneers such as Ole Fanger to the more progressive experts in the field of adaptive comfort at institutions such as the University of California–Berkley, many attempts have and are being made to refine what is known about what makes us comfortable or not.

This might, at first, seem not to be of direct interest to those people connected to the production of sustainable buildings, or the sustainable construction process, but it is vital. Unsurprisingly, the major truth that lies behind theories of thermal comfort is that a range of internal conditions suit a range of different people. The variances in our preferences for different temperatures, both air and mean radiant (normally sunshine), draughts (air speed), and quantities of moisture in the air (humidity) mean that there are a range of values, with generally accepted maximum and minimum restrictions that building design teams aim for. Several standards exist that can guide the reader in more detail, including ISO 7730:2005 (last reviewed 2009), *Ergonomics of the thermal environment: Analytical determination and interpretation of thermal comfort using calculation of the PMV and PPD indices and local thermal comfort criteria* (ISO 2005a); ISO 10551:1995 (last reviewed 2011), *Ergonomics of the thermal environment: Assessment*

3.1 Want to know more? Adaptive thermal comfort

An adaptive comfort approach makes allowance for the way in which people adapt to changing conditions in their environment (Nicol & Humphreys 2002). This adaption can be associated with different temperatures and relative humidity during different times of the year. The fact that an occupant may have experienced a hot summer's day could influence their perceived thermal comfort when entering an air-conditioned building. Research undertaken upon occupants of naturally ventilated buildings found that they accept, and in some instances prefer, a wider range of temperatures than do occupants of air-conditioned buildings, often due to the influence of outdoor conditions. This is particularly pertinent to the choice of acceptable ranges of temperature that can be assumed to be appropriate for buildings that do not use air-conditioning or mechanical ventilation.

The reader can assess the difference that varying temperatures and other variables make to predicted thermal comfort by using an online tool that undertakes thermal comfort calculations using the ANSI/ASHRAE Standard 55 (see Tyler *et al.* 2013).

of the influence of the thermal environment using subjective judgment scales (ISO 1995); and the ANSI/ASHRAE Standard 55-2013, *Thermal environmental conditions for human occupancy* (ASHRAE 2013).

One of the major advantages of designing a building using a comprehensive set of building services is that the internal thermal conditions of the building can be maintained within a small margin of error. If the designer decides not to use HVAC equipment to heat, cool, humidify, or filter the internal air, the internal spaces need to rely upon the building's fabric to regulate its internal environment. What aspects of the external and internal building fabric are of interest to a building designer?

3.3.3.1   Insulation

The use of insulation in buildings became prevalent in the UK in the late 1950s, and the need for insulation was standardised for all new buildings as part of the 1965 building regulations, part L (now entitled 'conservation of fuel and power'). In these regulations the minimum allowed insulation placed in the loft space above the ceiling of a traditional home was just over 25mm, or one inch, which equates to a thermal transmission or U-value of approximately 1.7 W/m²K.

As the need for better insulation was kick-started by the oil/energy crisis of the early 1970s, the regulations were amended to increase the values of insulation to reduce the quantity of energy used in homes (Table 3.8).

The accepted measurement of insulation is not thickness, for the obvious reason that the insulation properties of materials vary. The thermal transmission or U-value is often quoted by designers and manufacturers alike when referring to the amount of insulation provided for a complete building element. However, the value of the thermal conductivity of each separate building material is the building block that makes up the thermal conductivity of a complete wall, roof, floor, window, or door. This value

Table 3.8 Changes in U-values stipulated in historical versions of the UK building regulations

Year of Building Regulation?	U-Values (W/m²K)			
	Wall	Roof	Floor	Windows (and Doors)
1965	1.7	1.42	—	5.7
1974	1.0	0.60	—	5.7
1981	0.6	0.35	—	5.7
1990	0.45	0.25	0.45	5.7
1995	0.45	0.25	0.45	3.3
2002	0.35	0.20	0.25	2.0
2010 (new dwellings) 2013 fundamentally unchanged	0.30	0.20	0.25	2.0*

*Also limit on air permeability of 10.00 m³/h.m² at 50 Pa.
Sources: HM Government 2010; King 2007.

3.2 Want to know more? Thermal transmission

Thermal transmission values describe the amount of heat that will pass through a square metre **area** of a build's fabric for a one-degree difference between the two sides of that building element.

Thermal transmission is measured in watts of thermal energy transmitted for every degree centigrade or Kelvin through each metre squared of the part of the building being measured (W/m²K). Thermal conductivity is measured in watts of thermal energy transmitted for every degree centigrade or Kelvin through each metre **thickness** of the part of the building being measured (W/mK).

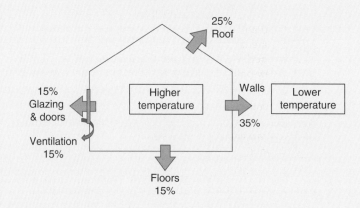

Figure 3.4 Thermal energy transmitted though the different elements of a building.

often depends upon the density of the material: the lower the density the higher the insulation value and the smaller the amount of heat transferred. This can be complimented by the linked nature of any air pockets or voids in a material, such as foamed insulants. If pockets of air contained within the insulant are interconnected, generally, the insulative properties of the material are less than an insulator of the same density whose air pockets are separate. A material that contains separate air pockets is normally termed as 'closed cell'. Table 3.9 shows the main types of insulation that are used in construction, where in the building you could find the insulation, and the thermal conductivity of each.

As walls represent a large percentage of the heat loss from a building and the number of possible design solutions for walls are numerous, Table 3.10 shows a range of different possible wall technologies with the broad specification for each underneath.

The solutions outlined in Table 3.11 are only focused upon meeting current UK building regulations. New sustainable buildings, particularly homes, normally aim at levels of thermal performance far superior to the suggestions taken from UK building control guidance (Camb 2014). This would lead to insulation that might be described as superinsulation.

Table 3.9 Typical insulations and where you might find these in a typical building

Thermal Insulation	Location of Insulation within a Typical Building	Thermal Conductivity of the Insulation
Mineral and glass fibres, both in more rigid 'bats' of rectangular cut sections or from a roll	Normally used in walls and roofs where the fibre materials would not be compressed and therefore allowed to maintain as low a density and high an insulative value as possible	0.035–0.040 W/mK*
Foams, rigid and pumped	Often in floors and wall elements, sometimes in commercial and industrial buildings, in roofs as well.	0.025–0.040 W/mK*
Manufactured natural materials, such as wood, wool slabs, straw-board	Can be used in most locations that suit fibre materials, and some of the more rigid products can be used under floors if appropriately protected	0.050–0.10 W/mK*
'Raw' natural insulation materials such as straw and sheep's wool	Many sustainable or deeper green buildings use these forms of insulation to reduce the overall impact of the building. Sheep's wool can offer a replacement for mineral fibre insulants, and straw bales can form complete wall sections.	Straw bale 0.055–0.065 W/mK (Sutton et al. 2011) (density 110–130 kg/m^3)
High-value insulation materials such as nano gels and aero gels	Many more insulants derived from improved engineering practices and new materials are gradually coming into the market. Translucent gels are one example of this and can be used in glazing to considerably improve the insulation of windows and roof lights.	0.013 W/mK (Thermablok 2013)

* http://people.bath.ac.uk/absmaw/BEnv1/properties.pdf

3.3.3.1.1 Superinsulation

This is a term commonly attributed to the insulation levels achieved in Scandinavian countries where the requirement for low environmental impact with high standards of comfort in low winter temperatures has led to the use of high levels of thermal insulation. One of the many attractions of increasing insulation standards is that the end result is almost maintenance free, requires little input from the householder, and can have welcome impacts on the internal comfort temperatures of buildings. The downsides can be that if the building detailing is not also improved to take account of this extra insulation, ingress of moisture from the elements or interstitial condensation can cause problems (Scandinavian winters often have much lower temperatures than a maritime climate such as the UK and, as such, have lower levels of external moisture in the air). Overheating can

Table 3.10 A sample of the range of methods that can be used to achieve a U-value of 0.30 W/m²K, for example

Wall technologies that will achieve a minimum U-Value of 0.30 W/m²K	Cavity wall—timber frame solution 1: 150 & 100×50 studs at 600 & 400 mm centres	Cavity wall—timber frame solution 2: 150 & 100×50 studs at 600 & 400 mm centres	Timber frame wall	Dry lining to existing solid wall	Full-fill cavity wall	Typical solid wall construction
Construction details	105 mmm brick, 50 mm clear cavity, 100 mm good quality fibreglass insulation, and 12.5 mm plasterboard	105 mmm brick, 50 mm clear cavity, 40 mm good quality insulating board, and 32.5 mm insulated drylining board	Tiles on battens external surface, over 100x50 mm stud wall filled with 100 mm fibreglass	215 solid brick wall lined with 68 mm insulated plasterboard and a 5 mm scim	105 mm brick outer skin, 100 mm glass wool–filled cavity, 100 mm lightweight thermally efficient aerated concrete internal block with drylining board	20 mm external render, 215 mm lightweight block lined internally with a 45 mm insulated dry lining

Table 3.11 Calculation of the U-value for a very 'standard' form of detail

Thermal Layer	Thickness of Material	Thermal Conductivity	Thermal Resistance
Thermal resistance outside	Not applicable	Not applicable	0.15 m²K/W
Facing brick	100 mm	1.0 W/mK	0.1 m²K/W (0.1m/1.0)
Cavity	50 mm	Not applicable	0.18 m²K/W
Insulating block	150 mm	0.12 W/mK	1.25 m²K/W(0.15m/0.120)
Thermal resistance inside	Not applicable	Not applicable	0.05 m²K/W
		Sum of resistance	1.73 m²K/W

also be an attendant issue if ventilation strategies are not also paired with superinsulation.

Many high-insulation, low-infiltration design philosophies and prefabricated buildings can be classed as superinsulated. The U-values associated with these buildings that qualify them as superinsulated buildings have varied over the years. As fuel prices have increased and the relationship between fossil fuel emissions and climate change has become more apparent, superinsulated buildings have had to increase their insulation standards to stay in front of the mandatory requirements as they have become

3.3   Want to know more? Thermal conductivity values and basic thermal transmission calculations

To establish the thermal conductivity values that are in turn used to calculate the U-values using the building materials in Tables 3.9–3.12, various testing houses across Europe and the globe use agreed measurement techniques and standards. Typically, thermal conductivity values were measured several times, often on 10 occasions or more, and the mean value was used to describe the thermal performance of a material. Since 2003 a European standard (harmonised product standards EN 13162 to EN 1317) states that instead of using a mean value, the thermal conductivity that can be officially cited and used in calculations relates to 90% of production within a 90% confidence level and requires a minimum of 10 heat flow meter or guarded hot plate tests. The quoted measured values also have to reflect the thermal performance of a product over a 25-year life span. This has, in effect, increased the quoted thermal conductivity values in product-related literature, which will in turn increase the calculated U-value and finally result in more insulation being needed to reach a claimed thermal performance (BBA 2012).

To calculate whether a wall, roof/ceiling, floor, window, or door meets current building regulations is relatively straightforward. However, to deal with every variation of possible façade would take too much time and be largely irrelevant as the basic premise of building up a thermal transmission or U-value is similar for all. So, why look at this? As U-values are quoted so frequently in connection with sustainable construction it is essential to tackle the basics. If more information is required, the approved documents of the UK building regulations, particularly Part L, Conservation of Fuel and Power (HM Government 2010), is a good reference. Other texts such as *Introduction to Architectural Science* by Szokolay (2004) and *Environmental Building Science* by McMullan (2012) are excellent sources of further information and methods.

The building blocks for U-values are the thermal conductivities that have been discussed earlier in the insulation section. The flow of heat that is resisted by a building element depends on the quantity and thermal conductivity of the materials it contains. This resistance will be further enhanced by the resistance of the inside and outside surfaces, along with any continuous cavities that are placed in the path of the flow of heat. Mathematically the surface and cavity resistances can be added together along with the sum of the thickness divided by the thermal conductivity of each of the materials. The final design U-value can be manipulated by altering the thermal conductivity values, the thicknesses, and the size and number of cavities in the building element. As most of the heat lost in many properties is through the walls (35% on average), it is logical to reduce the heat loss in this critical place to try to achieve the lowest U-value for the walls of a potential sustainable building. Below, Table 3.11 shows the calculation of the U-value.

For this wall, if the reciprocal of the sum of the thermal resistance is calculated, this will give the thermal transmission or U-value of the wall. In this case $1/1.73 = 0.578$ W/m^{2}K. Compared to the current building regulations requirement of 0.30 W/m^{2}K, insulation will be needed to reduce the design thermal transmission value of 0.578 to 0.30 W/m^{2}K. The U-value would, in reality, be influenced by the bridging effects of mortar joints. The ability to emit and absorb radiant heat, the emissivity, and the exposure ratings influence the surface resistances. More information concerning the details behind calculating a more realistic U-value than that shown in Table 3.11 can be found in building services guides (CIBSE 2006, ASHRAE 2013), government and research texts (the UK's BRE), and textbooks (Anderson 2006; McMullan 2012).

Figure 3.5 Homes at the UK BRE innovation park (Photo by Steve Goodhew).

more stringent. A superinsulated building of the 1970s would barely pass the thermal requirements of the current UK building regulations and probably use more energy than any of the homes built at the BRE innovation park (Figure 3.5).

As an example, compare the current UK regulations with typical superinsulation U-values. These can be written as simple contrasting figures with the thermal transmission (U-value) of each stated against the other: UK regulation U-value (W/m^{2}K) / typical superinsulation U-values (W/m^{2}K). For the wall the contrast is 0.3/0.15, floor 0.25/0.15, ceiling-roof 0.2/0.10, and for the glazing 2.0/1.2.

The law of diminishing returns applies to insulation and influences the amount of insulation that can sensibly be introduced into the fabric of any building. Factors such as the thickness of walls and, as previously mentioned, the potential to introduce overheating in summer will impede insulation levels; in addition, adding ever more insulation will provide incrementally smaller percentage improvements. As each doubling of insulation thickness takes place, less and less extra benefit will be received. This will be in the form of less energy saved but also more and more embodied energy locked up in the building's fabric. Superinsulation can only go so far. Figure 3.6 shows the relative reduction in theoretical energy cost for two insulation materials through summer and winter

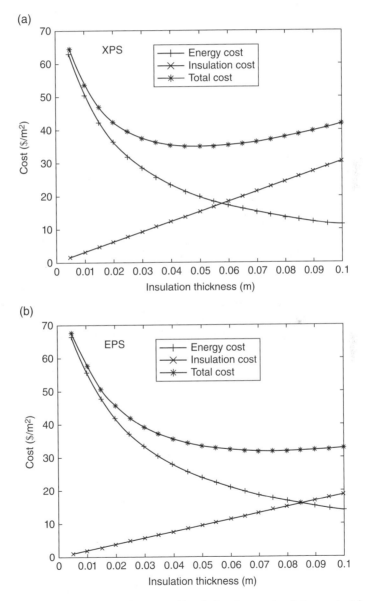

Figure 3.6 Energy use against the cost of insulation for two insulation materials over a 10-year period (From Ozel 2011).

conditions over a 10-year period compared to the insulation cost. As the cost of energy fluctuates, the optimum point will change (Ozel 2011).

A home built to Passivhaus standards (see Section 3.3.3.4.3 in this chapter for more details of Passivhaus) using cavity walls, the *Denby-Dale Passivhaus* built in 2010 has target U-values of less than 0.15 W/m²K for external envelope by using 300 mm fibreglass insulation in the wall cavity, 500 mm in the roof void, 225 mm in the ground floor, and a combined (frame and glazed unit) U-value of 0.8 W/m²K for the windows. A free technical briefing is available from the green building store (Green Building Store 2013).

Levels of insulation are only part of the story that influences the thermal performance of a building; the way a combination of materials allows heat to be stored is also important, alongside the ability for materials to reduce thermal transmission. As well as thermal conduction there is also thermal capacity: the ability of a material to store thermal energy.

3.3.3.2   Thermal mass

Thermal mass is a term that refers to the ability of all or parts of a building to store heat. This property of the building and its materials can influence how quickly the interior of the building will heat up or cool down. Materials that can effectively store heat often are denser than materials that store less heat. However, this trait is not followed by phase change materials, which store the thermal energy used to change their physical state (such as wax melting) as well as by storing heat through the increase in their temperature. The ability to store heat will normally rely on how many Joules of heat energy can be contained in a kilogram of material for every degree increase in temperature. This is termed specific heat capacity and is normally measured in Joules per kilograms per degree centigrade, J/kg K, (a rise or fall in degrees Kelvin [K] is equivalent to a fall or rise in centigrade [°C], but measurements in Kelvin are measured from absolute zero [−273.18°C as a datum]). The higher the specific heat capacity, the larger the quantity of energy that can theoretically be stored. Table 3.12 shows some of the commonly used materials associated with introducing thermal mass into a building.

Phase change materials deserve a separate mention as their use has increased due to their inclusion in various easily utilised board materials (BASF 2013). BASF have introduced a plasterboard with beads of wax built into the thickness of the board. The wax has a set temperature range when it will soften and eventually change from a solid to a liquid. The latent energy that is needed to enable this change is stored in the board until the temperature drops below the point when the wax will resolidify and the heat will be released as the solidification takes place. The temperature at which the wax changes state has to be matched to the temperature that the building design team believe will reduce the incidence of overheating or modify the internal environment in the building. The product can be selected so that the wax changes state over a

Table 3.12 Some of the commonly used materials that are acknowledged as able to effectively store heat in a building (figures from Goodhew & Griffiths 2005; CIBSE 2006, reproduced by permission of CIBSE)

Material or Class of Materials	Density (kg/m³)	Specific Heat Capacity (J/kg K)	Approximate Thermal Conductivity (W/mK)*
Dense concrete block	2300	1000	1.63
Lightweight concrete block	600	1000	0.19
Brickwork	1700	800	0.84
Water	1000	4200	0.60
Earth-based building materials (cob)	1450	800	0.45
Mineral fibre insulation	30	1000	0.035
Phenolic foam insulation board	30	1400	0.040

*Can vary according to internal structure and density.

series of defined temperature thresholds that might typically include 21°C, 23°C, and 26°C. At these temperatures the wax changes state and absorbs heat energy and stores it by changing phase. When the temperature falls under the temperature threshold, the capsule releases this stored heat energy again. The BASF house on the campus of Nottingham University uses this technology to guard against overheating. Further information related to phase-change materials, including organic and inorganic types and their application, can be found in the review by Baetens et al. (2010).

3.3.3.2.1 Position, quantity, and lag times of thermal mass materials

A major requirement when building sustainably is to reduce energy usage but keep comfort conditions within acceptable limits (see Section 3.3.3, Sustainable Heating and Cooling). Designing a building so that thermal mass is positioned to absorb unwanted heat by suitable materials (see Table 3.12) connected to exposed surfaces can lead to a cooler interior when compared to an identical building of low thermal mass. Some of the critical features that determine whether the mass will be effective are the following:

1. That the massive material is placed in an effective position. For example, having a heavyweight, tiled concrete floor in a sun space will absorb enough of the daytime solar radiation to moderate the space's internal temperatures.
2. That sufficient but not excessive depth of material is used. Thick walls of massive material will absorb heat into approximately the first 100 mm of the wall. Beyond this depth the thermal mass is unlikely to be beneficial.
3. That some form of ventilation can be utilised to transport the heat away from the inside of the building. If the thermal mass is being used to

moderate daytime temperatures, a series of hot days will 'fill' the capacity of the massive building, and without nighttime ventilation the building is still likely to overheat.

4. That the time lag offered by the thermal mass (the time between when the thermal energy is absorbed/stored within the material and the time when the heat is released) suits the characteristics of the heating and cooling needs of the building.

In order to moderate the internal temperature of a space, a material of high thermal storage capacity is introduced. The diagrams in Figure 3.7 show the process in an ideal situation applied to summer and winter conditions relying on diurnal (day/night) temperature differences.

It must be emphasised that the patterns of use, heating, quantities of internal gains, and general building use can assist or interfere with these ideal conditions. These aspects need to be taken into account when trying to maintain stable internal comfort conditions and when reducing energy usage.

3.3.3.2.2   An example of the use of thermal mass in buildings

More low-tech thermal mass solutions can be used: a rammed earth wall at a building at Hill Holt Wood provides a lecture theatre and an adjoining glazed space with high-mass surfaces suitable to absorb heat. Nottingham Trent University has undertaken a project placing thermocouples through the thickness of the wall (Figure 3.8) and at different heights and orientations in an attempt to measure the temperature change through a wall of this nature, in a real building.

3.4 Want to know more? Thermal mass, admittance, and thermal capacity

To calculate the impact of thermal mass in a building, the thermal admittance (a quantity that describes how the thermal transmittance of a material fluctuates with time, thereby allowing for the thermal capacity of the material, measured in the same unit as U-values; W/m^2K) of a structure needs to be calculated. The higher the admittance value related to a particular material, the better the material is suited to storing heat. As this is venturing into the area of relatively complex building physics, the reader can find out more through EN ISO 13786:2007, where thermal admittance is described in detail (ISO 2005b). Thermal admittance is also utilised in the CIBSE Simple Dynamic Model for calculating cooling loads and summertime space temperatures (CIBSE 2006). More complex solutions to questions relating to quantities of thermal mass and where to position that mass can also be investigated using dynamic thermal simulations, which are described from an energy point of view in this chapter and from the standpoint of assessing buildings in Chapter 6.

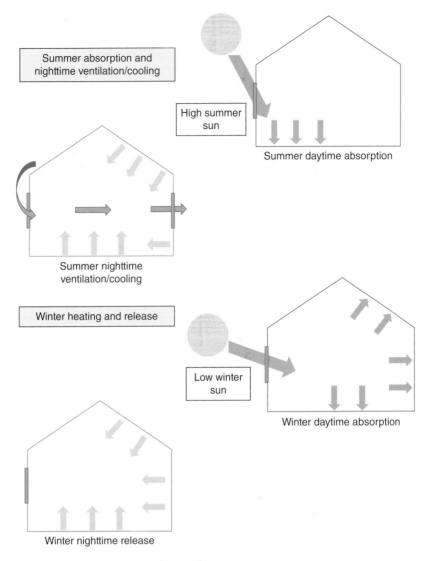

Figure 3.7 Summer cooling and winter heating.

3.3.3.3 Ventilation and energy efficiency

One of the major issues that have to be assessed alongside insulation and thermal mass is the need for adequate ventilation whilst maintaining good energy efficiency within buildings. The major objective of ventilation is to provide comfortable and healthy buildings. Achieving this, alongside reducing the amount of energy needed to heat or cool interior air, is relatively straightforward, but there are subtle issues to take into account.

3.3.3.3.1 Why do buildings need ventilation?
The ventilation of a building's internal spaces prevents some negative issues whilst providing some positive effects. Overventilating can sometimes challenge the maintenance of good internal comfort standards and

Figure 3.8 Thermocouples being placed into the rammed earth wall at Hill Holt Wood Centre, near Lincoln, UK (Photo by Steve Goodhew).

impacts the energy efficiency of the whole building. Underventilating increases the presence of stale moist air alongside associated risks of condensation and mould growth.

3.3.3.3.2   How are our buildings' internal spaces ventilated?
(Figures 3.9 and 3.10)
Background ventilation. Background ventilation is a low-level, controlled amount of air supplied to ensure good air quality and prevent any long-term buildup of water vapour without causing draughts or compromising the security of the building. It can be termed infiltration or trickled ventilation. Often, older buildings have enough gaps left under doors and between suspended floors, as well as air entering via chimneys, to supply enough, or even oversupply, background ventilation.

Ventilation can be needed for set purposes such as an air supply for a combustion device, a boiler, a wood burner, or perhaps an open fire. A routed air supply might not enter the building in the case of a balanced flue. Reducing this form of ventilation is very unwise, dangerous, and can be illegal.

Purge ventilation. Another specific but short-term ventilation is termed purge ventilation. This might be natural or mechanical and is sometimes as

lead to a building having a reduction in energy demand compared with the average energy demand of dwelling that will pass the current building regulations. Currently this equates to an energy usage of 15 kWh/m² per annum as compared with a dwelling built to current building regulations using 200 kWh/m² per annum. This level of energy usage equates to a quantity that can be supplied by renewable sources, such as those discussed in Section 3.3 of this chapter. This is achieved by using five main principles, shown in Table 3.14.

The below priorities describe the ways in which passivhaus dwellings achieve their goals, but they do not give the full story. There is also the value of passivhaus dwellings when related to the home owner and beyond. Other relevant aspects are:

- The dwelling should be affordable to buy/rent and run, offering true value to the occupants.
- The dwelling should be of a quality that will delight any occupant and ensure long operation times before any retrofit is required.
- Emissions from any dwelling should be appropriately low.
- A high-quality internal space should be offered with appropriate day lighting and above-average air quality.

Table 3.14 Five main principles behind a Passivhaus design

Passivhaus Imperative	Detail of Imperative
1. A barrier to heat loss	Thermal transmission (or U) values of all building components to be as low as is currently feasible. For example, U-value of walls to be equal to or less than 0.097 W/m²K.
2. Heat recovery	To use a ventilation system (MVHR) that recovers up to 80% of the heat from the air in the dwelling by passing the warm, moist internal air through a heat exchanger, transferring the heat to the cooler, drier incoming air. This will not only save energy but also provide high levels of air quality, leading to equally high levels of thermal comfort and less potential degradation of the internal finishes in the home.
3. Heat-collecting glazing	By using glazing systems that have very low thermal transmission values (see Imperative 1), the orientation of the glazed surfaces can used to trap solar gains and rightfully expect that the glazing will not act as a thermal weak point in the dwelling's façade.
4. Junctions without thermal bridges	To allow the insulation of the façade, floors, and roof to be continuous and not allow heat to escape through poorly detailed junctions.
5. Airtight construction	The normal supply air requirement of a dwelling is 30 m³/person (BPIE 2015). However, the typical dwelling fabric allows 120 m³/hr into the home. If the building fabric is detailed and built to high standards, with few gaps, good joints, and well-fitting doors and windows, the main air supply to the home can be through the heat recovery system with the benefits discussed in Imperative 2 of this table.

The test of these claims comes when passivhaus buildings have been monitored and occupants surveyed to ascertain whether the performance of these buildings do in fact match the design-related assumptions. Post-occupancy evaluation (see Chapter 5) has been undertaken in a large number of passivhaus dwellings and compared with other low-energy houses. The first passivhaus that was built in Darmstat has measured results for 16 years from when it was occupied in 1991. The average annual energy consumption was 25 kWh/m², close to the target figure (bearing in mind this was the first of this building type). Local authority, housing association, and public buildings are now being built across the globe to passivhaus standards; however, passivhaus standards are not applicable to all buildings. The sealed nature of the interior environment demanded by passivhaus does not necessarily sit well with breathable building designs or with materials that are more porous and organic in nature. Refurbishments of some buildings can meet the stringent thermal and infiltration demands of passivhaus, but many cannot. The reduction in interior footprint due to extensive thermal insulation upgrades is one major reason for this. However, if the fabric of a building due to be refurbished can meet the low infiltration standards and has the scope for thermal upgrading, passivhaus may be one avenue to take to improve energy use, quality of environment, and long-term affordability.

3.3.3.4.4 Breathing buildings

Breathing buildings are a part of specific design philosophy that links the use of certain materials, many of which are obtained directly from natural origins, to provide a path for the diffusion of water vapour through the building's fabric. The rationale for this approach comes from a number of different drivers, but the major one involves the inability of many buildings to maintain a vapour-resistant layer sufficiently intact to prevent moisture from entering into the depths of a building element, such as a wall or ceiling/roof. Typical examples of this are the paths left by the electricians and plumbers who need to penetrate any vapour checks. This check is normally formed by a suitable-thickness polythene or other sheet which needs to be taped around a pipe or cable whenever the layer is cut. In reality, this is a difficult process to complete under site conditions, and therefore moisture can in some instances migrate into the centre of an insulated exterior wall or ceiling. This moisture could promote deterioration of the building fabric or be a catalyst for the growth of mould inside the building element alongside poor indoor air quality. Breathing buildings, through the vapour-permeable materials chosen for their walls and ceilings, allow water vapour to diffuse through to the exterior. As most buildings in the winter months have reduced ventilation levels—and the interior temperature is higher than the exterior temperature—the interior space will be at a pressure higher than the exterior. This provides the mechanism that will enable water vapour to pass through the buildings' external fabric as well as a reduction in the overall environmental impact of a building.

Another aspect of breathing buildings is the use of materials that have a high degree of hygroscopicity, the ability for a material to absorb moisture. This can allow the surfaces and sometimes the internal layers of a wall or ceiling to absorb moisture, reducing the peaks and troughs of internal moisture content. This in turn can help prevent the increase of internal relative humility to over 70%, which will also reduce the possible incidence of internal surface condensation.

A number of researchers have looked into the efficacy of breathing solutions, walls being the most prevalent target. John Straube (Straube & de Graauw, 2001) has undertaken some practical measurements and concluded that breathing walls can moderate indoor humidity and therefore practically illuminate the potential for fungal growth on building services. Simonson (Simonson *et al.* 2004) found that the impact of the use of hygroscopic materials in a building in a Canadian climate did increase the time that the indoor air is too dry (RH < 25%) but decreased the time that the air is too humid (RH > 60%). Through an associated investigation it was found that hygroscopic materials typically improve warm respiratory comfort and air quality conditions during occupied hours. Whilst breathing solutions are not always possible, such as where an existing substrate is impermeable or other materials properties reduce materials choice, where appropriate, this form of specification can work and help provide good indoor air quality.

3.3.3.4.5 High insulation and thermal mass: BedZED

The Beddington zero energy development (BedZED) is an urban development designed by the architect Bill Dunster, which was/is the UKs largest sustainable mixed-use community development. This project used many techniques that have been employed elsewhere, but never as boldly or so holistically. Economic and social aspects of the philosophy come to the fore, with shared housing for key workers mixed with private homes. The location of the development was seen at the design stage as an important issue (combined with the sharing of low-emission vehicles and access to public transport). The building technology philosophy linked high thermal mass to high standards of thermal insulation, renewables, district heating (originally from biofuels, but now from gas), and the famous chicken head ventilators (Figure 3.11).

The BedZED project has a major advantage: the performance of the whole development has been monitored from 2002 through to the latest Bioregional published survey in 2007. One issue with some sustainable designs is the lack of proof that their original principles do in fact work when the building or structure is in use. The issue of monitoring and the more broadly framed practice of post-occupancy evaluation is discussed during the construction phase chapter, Chapter 5. The results from Beddington's monitoring can then be compared with national norms in the many areas that are viewed as sustainable indicators (these indicators are discussed in Chapter 6), shown in Table 3.15.

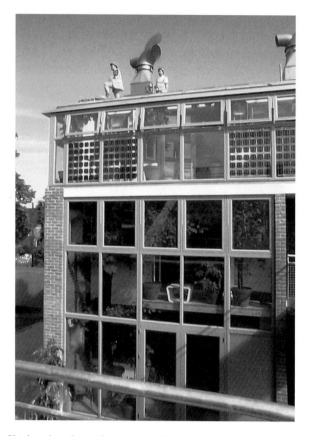

Figure 3.11 Chicken head ventilator at BedZED. Reproduced by permission of Brian Pilkington.

Table 3.15 Monitoring results from Beddington's zero energy development

Sustainable Indicator	UK National Norm	BedZED Value
Water use	Over twice BedZED	72 litres/day avg.
Electricity use	BZ uses 42% less than Sutton (the locality)	2.579 kWh
Heat (from gas)	BZ uses 81% less than Sutton	3.526 kWh
Carbon footprint	BZ uses 4.67 global hectares	5.6 global hectares

Having monitored BedZED, Bioregional stated that if the combined heat and power system was fueled as intended, the overall carbon footprint would be less than the 4.67 global hectares and closer to 4.32 global hectares (ecological footprinting is described in more detail in Chapter 6). However, according to BedZED this would still leave the BedZED residents 1.32 hectares short of the three hectares needed to live within the confines of one planet's worth of usable land. WWF indicates that, presently, the global hectare requirement is 1.8, so there could be some debate as to how close BedZED has come to meeting its target. In defense of the design, the goals that were set as part of the initial concept were ambitious and most were achieved. The whole project showed that those elements that reduced the overall impact were outside the control of the designers. For example, designers are unable to influence carbon emissions associated with food intake, the resident's workplace, and services such as schooling.

The Zero Bills home comes from the same stable as BedZED. Through a combination of affordability and renewable generation, it offers a home that generates electrical energy for eight months of the year, offsetting four months of winter when the building requires electricity from other sources (Zero Bills Home Company 2013).

3.3.3.4.6 Highly insulated tested building fabric: Super E

A Canadian system (supported by the Canadian government) built in a number of countries of low energy construction, termed Super E, relies on a series of individual tests to ensure that the completed building has low air permeability with appropriate ventilation alongside modeled energy consumption benchmarked against ideal standards. There is now a Super E Net Zero Energy Home, which is heated by solar heat, internal gains (cooking, lighting, people, and appliances), and by heat extracted from the ground by a ground source heat pump. As the built property is in Japan, there is a requirement for cooling, so there is a combination of shading and natural ventilation augmented by mechanical cooling from the ground-source heat pump.

3.3.3.4.7 High thermal mass with glazed southern façade: Hockerton Housing Project/Autonomous House

Whilst the Hockerton housing project obtained planning permission in 1996, the principles behind the project (described in the book *The Autonomous House*, 1975, and updated in *The New Autonomous House*, 2000; Vale and Vale 1975, 2000) are valid today. The project architects, Robert and Brenda Vale, advocate an off-grid solution that utilises a combination of high levels of both thermal mass and insulation in the building's fabric. This is further augmented by earth-berming the rear of the five properties and glazing the south-facing front façade. As the heating and cooling requirements are so low, much, if not all, of the energy needs can be supplied by relatively modest site-based renewable energy supplies.

To allow more detailed investigations into the impact of adopting any of these or other low-energy design philosophies, it is very useful to be able to model the consequences of these design-/energy-related decisions. Building performance simulation allows the ultimate flexibility of trying different energy strategies when deciding to finalise either an early stage or more fully formed building design. The next section describes the basic processes used by building performance simulation, and Chapter 6 contains a further section that shows how building simulation can be used as an assessment tool.

3.3.3.5 Building performance simulation

Building performance simulation has been used for many years either to undertake simple assessments at an early design stage or to hone a design proposal. The original simulations worked on large mainframe or workstation computers, often taking hours or even days to calculate the performance of a building.

Simulations are now undertaken using PC- or MAC-based software, and the complexity of the package is normally matched by the complexity or amount of information required to undertake building performance measurements. The categories of information normally used to build up a simulation model are shown in Table 3.16.

As so much information can be captured by a single CAD drawing, most commercially available simulation packages use a graphical 'front end' or interface that will allow a drawn building to provide much of the basic dimensional information for the software. The use of Building Information Modeling (BIM) will increase the linkages between different information sets, and many simulation package suppliers are looking at the opportunity to link a number of data sets together in one powerful model. The levels of BIM and the possible impacts of this data sharing are discussed in Chapter 6. CAD drawings produced using other packages can often be modified to be absorbed by the simulation package, making the task of preparing the

Table 3.16 The categories of information used to build up a simulation model

Information Needed to Simulate a Building's Performance	Examples of the Information in Each of the Categories
Building dimensions	An obvious requirement, but accurate dimensions are needed to calculate basic properties such as thickness of walls, quantities of air in rooms, size of glazing.
Building materials, services, and building data	The thermal and associated properties such as densities need to be known to allow appropriate analysis of the building's internal comfort conditions in winter and summer. The building services need to be defined along with rates of infiltration and relative humidity.
Occupant behaviour	Occupant behaviour patterns such as workspace usage patterns: when do the cleaners switch on the lights and require the heating to be on in the winter months? The number and location of people in a building will influence the amount of cooling that might be needed in the summer months
Weather patterns	The geographical position of the building to be simulated will require that appropriate averages of temperatures, sunlight, precipitation, and other weather data of that locality are used to accurately simulate the building's performance.
Patterns of building services use	The basic on-off patterns for heating and cooling and also the added issues of lighting and cooking

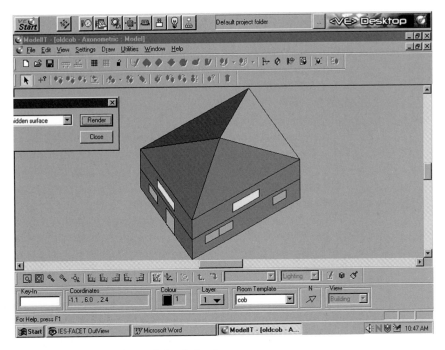

Figure 3.12 Dynamic thermal simulation input screen.

simulation model considerably easier. A typical graphical representation of a potential building is shown in Figure 3.12.

Of the many packages that are available, several well-known systems have been produced wholly or partly in the UK; these include TAS, IES, and ESPR. These packages will take the information that the design team has put into the software and enable the team to investigate different possible solutions and scenarios. As with most software, an obligatory 'health warning' is necessary. 'Garbage in, garbage out' is a well-worn but appropriate statement that alerts the software user to only trust the results of the calculations as much as they trust the data and assumptions that were used to create the simulation. However, the opportunity to try out ideas before the building has been constructed and undertake some simple 'tweaking' to a design is one that should not be missed if the building's type, size, or complexity warrants this. Building simulation can also be used to assess the performance of a potential building and compare the results with those of designs that might be termed 'best practice'. This function is further developed in Chapter 6. One way in which the performance of a building can be made more sustainable is the inclusion of cleaner renewable energy generation into the building design. Many simulations will now allow a range of typical types of renewable generation to be modelled, and the results can be compared with the same model but without the inclusion of the renewables. The next sections examine the options for cleaner energy generation and reducing energy consumption as part of the design of a sustainable building or structure.

3.4   Energy and design: Renewable energy and sustainable technologies

This section is focused upon the renewable options that might be chosen when building sustainably. Chapter 8 also offers a different slant on renewables, concentrating upon the issues that occur when installing or running such forms of electrical and heat generation.

3.4.1   Renewable energy and the design of sustainable buildings

Renewable energy can be defined as energy coming from natural sources, such as from the sun, wind, and tides, and are therefore categorised as being naturally replenished (i.e., renewable). Some forms of energy generation have been categorised as semi-renewable or hybrid systems; these might use an efficient generation system partially reliant upon waste heat, such as combined heat and power, or use a bank of heat from the surroundings, such as heat pump systems.

The European heads of state or governments agreed in March 2007 on binding targets to increase the share of renewable energy within the total energy generated in Europe. By 2020 renewable energy should account for 20% of the EU's final energy consumption (EU 2014a) (8.5% in 2005). The UK's current percentage of renewable energy generation is relatively small but increasing, and through the EU agreement should be 15% by 2020 (EU 2014b). Bridging this gap might be achieved using general renewable sources of energy, such as tidal barrages, offshore wind, and wave power. However, it is possible to use buildings as a platform for placing renewable and semi-renewable sources of energy. This section examines in turn each of the separate renewable and semi-renewable or hybrid sources and their relationship to sustainable construction.

3.4.2   Renewables: Solar-related heating and photovoltaics (PV)

The gain of solar energy within buildings can be characterised as either passive or active in nature. Active heat or energy gain is received through a technological system. Passive gain is achieved through more traditional means, like solar gain glazing and façade orientation. Passive systems are dealt with in Section 3.3.3.4.

Active systems can theoretically be added to buildings that have appropriate orientation to augment the use of glazing to trap solar heat. Although solar energy can be harnessed in many ways, including specialist systems that can heat air via the cavity of a wall and cool a building using solar absorption and desiccants, there are two main energy collection technologies used in this arena: heat (solar thermal) and electricity (PV). Further information can be found in a variety of publications, including BSRIA's *Illustrated Guide to Renewable Technologies* (Pennycook 2008). Both technologies have their advantages and drawbacks.

3.4.2.1   Solar thermal (often termed solar hot water)

Solar radiation will heat up any object that is placed directly in the sun's rays. The major elements that affect the amount of heat absorbed by any object include the angle and azimuth of the surface of the absorber, the insulation below the surface, and any glazing/evacuation over the surface. For a solar thermal array or photoelectric panel to perform at optimum efficiency, the array, panel, or tubes need to be at right angles to the sun's rays. This minimizes the area of collection or maximizes the energy density. London lies on a latitude of 52 degrees; therefore, as the equinox (mid-March and September, equal day and night) is reached the sun will be 90 minus 52 degrees above the horizon, at 38 degrees. In the summer months the sun is higher in the sky and therefore the panels would be best placed at about 60 degrees (mid-June) from the vertical. It would be unusual to optimise winter conditions, but if these were a priority the lower sun would require panels to be placed close to 14 degrees (mid-December) as shown in Figure 3.13. If the reader wishes to experiment with different locations on the earth and times of year, there are several online solar angle calculators, including one from *The Solar Electricity Handbook* (Boxwell 2013).

To maximize the quantity of thermal energy that is trapped by a solar thermal system, the technology needs to effectively absorb and retain the heat from the sun. Two solar thermal systems, flat plate and evacuated tube systems, are commonly used on buildings. Both use slightly different ways to insulate the hot fluid that flows through them from the colder outside air. Flat plate collectors are the simpler of the two systems, are often the least costly, and are shown in Figure 3.14.

A flat plate, normally of a conductive metal such as copper, has a loop of pipework, often also made from formed copper sprayed with a coating designed to enhance the absorption of radiant heat of the surface. The plate is covered by glass to trap the long-wave infrared radiation between the glass and the inner surface and build up the temperature. The plate is insulated underneath and an equally well-insulated water store is linked to the plate(s) by a flow of fluid that moves the heat from the plate(s) to the water storage vessel. The more expensive alternative system uses conducting U tubes placed into an evacuated glass tube open at one end to allow the double skin to wrap around the U tube in place of the flat plates (Figure 3.15).

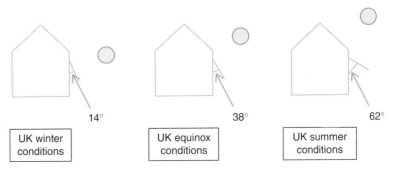

Figure 3.13 The optimum angles for solar generation in London, UK.

Figure 3.14 Flat plate collectors solar thermal system (Photo by Steve Goodhew).

Figure 3.15 Evacuated tube collector solar thermal system (Photo by Steve Goodhew).

The advantage of the evacuated tube solar thermal systems is their efficiency. The evacuated tube directs more sunlight to the internal U tube and provides enhanced insulation against heat losses from the circulating fluid. The amount of heat that can be trapped by evacuated tube systems is generally too great for water to be safely used and types of oil or other liquids are often used to transport the heat from the tubes to the storage vessel. This necessarily involves the use of more complex heat exchangers whose effectiveness, aligned to the size and type of water storage included in any system, have to be optimised for good overall efficiency. Both evacuated tube and flat plate systems normally have to use a small pump that circulates the transfer fluid. In some instances this pump can be powered by a small PV panel; thus, when the sun is shining most, the pump will circulate the fluid at the greatest extent, whilst at night the pump will not be operating, shutting the system down when not needed. Some very simple flat plate systems can operate without a pump, and this can sometimes be seen on the flat roofs of southern Mediterranean countries. The flat plate angles towards south and a water storage tank is placed upon a stand raising it above the upper level of the flat plate, using quite large diameter pipes. The heated water will rise up the pipe connected to the tank from the upper part of the plate. Cooler water will flow down from the tank to the lowest point on the flat plate initiating a convective or buoyant flow of heated water to the higher storage tank. The storage tank itself can be used as an absorber, albeit an inefficient one, reducing any losses from the system. In hotter climates the overall heat loss of the system will be lessened by the high outside temperatures.

Solar thermal installations are viewed as one of the most cost-effective renewable energy technologies that can be installed. The total capacity of solar thermal heating in the UK was estimated to be 73,640 kWh in 2010, an increase of 18.1% over the year and likely to continue increasing as UK government incentives come into play (ESTIF 2011).

The cost of solar thermal systems often will depend upon the likely payback period, which in turn is influenced by the expected heat output and the capital cost of the installed system. The Energy Saving Trust undertook a field trial of 54 flat plate installations and 34 evacuated tube systems, comparing the different formats alongside different pumping, heat exchange, and storage options (Energy Saving Trust 2011).

Overall, the study found that between £100 and £150 per year was being saved by the households. Surprisingly, despite a prediction by the UK's Energy Commitment Scheme that savings in the UK from flat plate collectors would be smaller (454 kWh/m^2) than from evacuated tube systems (582 kWh/m^2), the Energy Saving Trust field trial found that the two systems were very similar in output: 'The median for flat-plate collectors was 1,156 kWh per year and the median for evacuated-tubes 1,140 kWh per year' (Energy Saving Trust 2011). Whilst the field trial only included 88 monitored systems, these findings do suggest that either system is a genuine option for UK buildings.

3.4.2.2 Solar electrical

Photovoltaic or solar cell panels have existed for many years, and their development was accelerated by a number of factors, including the need for reliable energy generation in one of the ultimate 'stand-alone' locations, outer space. The panels themselves use a range of similar materials to convert the sun's rays to direct current electricity. The type of material used in the panel has a direct influence upon the efficiency of that conversion. Amorphous silicon films are the cheapest but least efficient of the panel materials, normally able to convert between 6% and 7% of the sun's energy to electricity. Greater efficiencies can be obtained by using

Figure 3.16 Ground, roof, and slate/tile mounted PV panels, Zurich Airport solar canopy (Photo by Steve Goodhew).

polycrystalline (9%–12%) or mono-crystalline (11%–15%) materials. As shown in Figure 3.16, the panels can be mounted at ground level outside of a building, on a roof as a solar slate or tile, or as building-integrated PV systems taking the place of a normal roof covering or façade material.

The average maximum solar radiation intensity incident on a building depends on where the site is on the planet. It depends on the site latitude and height above sea level. Manual calculations can be used to estimate the output of a PV array, but a number of online, or bespoke, calculators can make the task much easier. The US National Renewable Energy Laboratory has developed such a calculator, PVWATTS, to quickly obtain performance estimates for grid-connected PV systems. Whilst several versions are available, the simple Version 1 allows a quick view of the possible performance of PV systems across the world (Renewable Resource Data Center 2013).

If a 4 kW peak (*peak* refers to the maximum output from a PV array in the best conditions) fixed-roof mounted system is modeled at various locations, the results from the calculator can be seen in Table 3.17.

More accurate software can be used to predict the performance of a PV system, including some free online versions such as that on the Canadian government–related website RETScreen (Natural Resources Canada 2013).

Features that can direct the design of some aspects of a PV array, beyond that of size, are PV panel type, array pitch, or predicted output; these relate to the availability of an electrical grid connection. The lack of mains electricity normally constitutes part of the attractiveness of a PV system. The obvious issue is the need for a method of storing the generated electricity for use at night. Other clients might want the convenience of an off-grid supply due to unreliable mains connections or the ability to be

Table 3.17 The AC electrical outputs from a number of 4 kW/P PV arrays placed in different places on the globe, estimated using the PVWATTS version 1 calculator (Renewable Resource Data Center 2013)

City	Latitude	Elevation	Array Tilt	Insolation (kWh/m²/day)	AC, Energy (kWh)
St. Paul Island, Alaska, USA	57.15° N	7 m	57.2°	2.44	2666
Aberdeen, Scotland, UK	57.20° N	65 m	57.2°	2.62	2764
Moscow, Russia	55.75° N	156 m	55.8°	2.98	3067
London, Gatwick, UK	51.15° N	62 m	52.3°	3.07	3163
Channel Islands, UK	49.22° N	84 m	49.2°	3.58	3812
Vancouver, Canada	49.18° N	3 m	49.2°	3.82	3950
Christchurch, New Zealand	43.48° S	34 m	43.5°	4.20	4467
Sao Paulò Congonhas, Brazil	23.62° S	803 m	23.6°	4.90	4970
Beijing, China	39.93° N	55 m	39.9°	4.87	5150
Sydney, Aus	33.95° S	3 m	34.0°	5.04	5212
Accra, Ghana	5.60° N	69 m	5.6°	5.22	5262
New Delhi, India	28.58° N	216 m	28.6°	6.02	5915
Addis AbabaòBole, Ethiopia	8.98° N	2355 m	9.0°	5.71	6030
Las Vegas, USA	36.08° N	664 m	36.1°	6.60	6654

Table 3.18 Characteristics of stand-alone and grid-connected PV systems

Characteristic	Stand-Alone	Grid Connected
Best suited situation	Isolated positions	With good connections to mains electricity
Continuity of supply	Continuity of supply not essential	Uninterruptible supply essential
Power requirements	Lower load categories	Higher load categories
Generation capacity of array	Lower generation capacity close to or under base-load required	Higher generation capacity over base-load required

independent. The differences in approach and installed technology between designing a stand-alone or a grid-connected system are not large but do have some nuances, described in Table 3.18.

A trait in common with solar thermal panels, or tubes, is that the peak efficiency depends upon the amount of direct sunlight the panel is capturing, the time of year (the higher the sun is in the sky and the longer it shines will increase the amount of electricity generated per day), the geometry of the panel, and whether the site (often the building's roof) is overshadowed or obscured. One element associated with PVs is the need for electrical storage if the panels are not connected to the national electrical grid. This and other more operational and construction-related issues are discussed in the construction section, Chapter 8.

3.4.3 Renewables (semi): Fuel cells and hydrogen

Fuel cells are a method of generating electricity from the combination of different chemicals, often gaseous elements such oxygen and hydrogen, generating water and heat as direct by-products. Fuel cells can be viewed as a form of battery with a constantly maintained delivery of fuels and therefore can carry on operating when normal battery systems would be spent. The fuel cell comprises an anode and cathode in an electrolyte and produces a direct current, much as photovoltaic cells. Thus, an inverter is required to convert the DC electrical output to alternating current. For sustainable buildings, fuel cells offer relatively high efficiencies across a wide range of electrical loads, allowing for the fluctuating electrical needs of a typical building. Each fuel cell only produces a modest voltage, less than 1 volt, so to supply the needs of a building a number of cells need to be linked into a stack. This modular system links a number of fuel cells and allows a fuel cell system to be operated at different levels, whilst also providing the ability to up-size and down-size. The fuel needs to be supplied to the fuel cell for it to function, and so fuel needs to be delivered either to site or generated on site. For this reason fuel cells cannot be viewed as an easily established 'stand-alone' technology, and the inherent advantages, disadvantages, and efficiencies of the equipment necessary to support a fuel cell system need to be factored into the sustainability and usability of the whole package.

A typical fuel cell system will use the following auxiliary components or system parts:

- a reformer (produces hydrogen from an appropriate fuel)
- a desulphuriser (removes sulphur, viewed as an impurity in this instance, from any fuel gas supplied to a fuel cell)
- conversion equipment to take DC to AC
- a water management system
- a heat transfer system
- a system to control the function of the fuel cells and peripherals

The major reduction in overall CO_2 emissions associated with fuel cells partly relies on the fact that the cells do not have a combustion stage associated with most other aspects of electrical or heat generation and the efficiency is not limited to those achieved by heat engines. As well as reduced CO_2 emissions, they also produce lower emissions of oxides of nitrogen, hydrocarbons, and particulates than most forms of combustion-based engine. However, the making of O_2 and H_2 for the cells can itself create CO_2.

3.6 Want to know more? Efficiency of fuel cell systems

There are many types of fuel cells, differing in size, technology, purpose, and cost. Polymer exchange membrane fuel cells are being supported by some government agencies and some research organisations, as this technology works at lower temperatures than many other systems and looks promising as a method to power homes and vehicles. Inverters also have an influence on the efficiency of the whole fuel system, as can be seen in Table 3.19.

Table 3.19 Efficiency of fuel cell systems (US Department of Energy 2011)

Fuel Cell System Component	Efficiency Range
Polymer electrolyte membrane (PEM)	35%–60%
Alkaline (AFC)	60%
Phosphoric Acid (PAFC)	40%
Molten carbonate (MCFC)	45%–50%
PV panel	
Monocrystalline	15%–18%
Polycrystalline	13%–16%
Hydrolyser	50%–80%
Inverter	75%–90%

One of the most efficient uses of fuel cells is as part of an extended combined heat and power (CHP) system (see Section 3.4.7 for more information concerning CHP), where the heat generated by the stack of cells is channelled into meeting the heat-related demands of a building. This will obviously increase the efficiency of the whole system, but the exact gain will depend on the heat produced by the cells. Different cell types operate at widely differing temperatures. Moreover, the temperature and flow rate required by the building will need to be phased to the heat production.

A number of helpful publications and contacts can assist with further information related to fuel cells; in particular, some helpful datasheets have been produced by CHP Group of the Chartered Institution of Building Services Engineers (CIBSE), www.cibse.org/chp, and the US Department of Energy's Fuel Cell Technologies Office (CIBSE 2013; US Department of Energy 2013; CIBSE 2014).

3.4.4 Renewables: Wind

Wind renewables can be used to produce mechanical or electrical power. Wind-powered mechanical devices have traditionally driven flour mills (Figure 3.17) or, in more modern times, pumped water.

By removing a mill or pump it is possible to drive an electrical generator, normally an alternator, to produce AC current. The basic requirements of traditional windmills still hold true for the more modern wind generators: appropriately sized blades for the alternator, enough wind to turn the blades, consistent wind source, and systems that can stop the generator 'overrevving' and damaging the generator or the blades. Various different generator configurations exist, and some function best at different locations, others for different varieties of electrical generator. The siting of the generator is important as this helps to satisfy many of the variables. The Building Research Establishment has undertaken an investigation of the efficiency of a number of domestic mounted wind generators. The results suggest that domestic wind generators can supply a small percentage of

Figure 3.17 Greek windmill (Photo by Will Goodhew).

the quoted outputs due to variable or low wind speeds; thus, wind generators are best used away from buildings. Similar requirements also stand for solar electrical systems: the need for a grid connection and/or method of electrical storage.

The generation of electricity from the wind can be a more delicate proposition than some other technologies, such as photovoltaics, as the conversion of kinetic energy from the wind requires moving parts, and these will produce associated vibration and some noise.

3.4.5   Renewables: Biomass

Biomass heating systems for buildings could be described as the earliest form of cooking and heating. For centuries wood and vegetable matter have heated, illuminated, and provided radiant and convective cooking in the form of spit roasting and, later, cooking in containers. Over 200 years ago farmhouses in the UK reputedly used their farm animals as a form of heat for the occupants. The major driver for the use of biomass is the smaller amounts of energy used to obtain the fuel, as the carbon that forms

have to be wasted unless a large and expensive storage system were installed. The matching of these base loads is often critical to the efficient use of any CHP system.

Recently, small-scale CHP installations have been coming onto the market, often referred to as micro-CHP units. These have a heat output close to that of a normal domestic boiler and an electrical output of between 1 and 3 kW. They are small enough to fit under the work surface of a kitchen and the technologies they use (often a Stirling engine) are quiet enough to allow them to be incorporated into the surroundings of a household. The prevalent technologies currently associated with micro-CHP are the following:

1. *Stirling engine* The Stirling engine is essentially an external combustion engine that is sealed and takes heat from its surroundings to expand a gas inside a cylinder, which in turn exerts a force upon a piston to create rotary motion that will turn an electrical generator. This form of engine is very quiet and therefore very suitable for domestic operation. However, the amount of electrical power generated compared to the heat output is approximately 1:6 in favour of the heat energy.
2. *Fuel cells* As described in Section 3.4.3, fuel cells have no moving parts so they involve very little noise or vibration, making them very suitable for the domestic market. Recent mass-market versions have been produced by a well-known domestic boiler maker, and these promise to supply 100% of the heating and 75% of the electrical requirements of a home.

The general advantages of micro-CHP:

- Can use a wide range of fuels, including (with some special measures) biofuels
- Possible income from export of electricity
- Provides security of reasonably priced supply of electricity
- Carbon savings of 5%–20% can be achieved in most buildings
- Can generate electricity at peak demand

The general disadvantages of micro-CHP:

- Capital and installation costs can be more than providing mains electricity feed alongside more traditional gas boilers, but this should reduce as economies of scale improve. Current payback periods can be as long as 20 years, which is obviously a disincentive related to a unit that may have a lifetime of less than this time period.
- Smaller CHP systems often provide smaller savings in carbon emissions, especially true if the carbon linked to grid-related emissions is reduced. Inefficiencies apparent in micro-CHP units are emphasised if the units are used in short cycles.
- The available supplier and installer network is currently limited.

- Micro-CHP is a new technology compared to gas boilers, and therefore, there are some implicit small risks associated with the reliability of these technologies.
- The issues of base load, particularly the use for the heated water, remain, particularly in the summer months.

The previous articulated advantages and disadvantages of micro-CHP do guide the potential installer to choose slightly larger units, minimize short cycles, and optimise other power and heat systems to make best use of the advantages of CHP.

These technologies are a welcome addition to the diversity of energy-efficient appliances and techniques used for generating low-carbon electricity and heat.

3.4.8 Renewable and sustainable technologies for historic buildings

Historic buildings with technological and design-related importance are crucial to our built environment. They are quite rightly protected by legislation that prohibits a number of categories of changes to their exterior and interior. However, older buildings are often constructed of materials and use designs that were fit for purpose in their day but now can be difficult to heat or cool, and can have high energy-related operating costs. Interventions are not permitted by legislation and can have detrimental consequences upon the materials and fabric of the building as a whole. For example, adding impermeable insulation to some historic building fabric can introduce thermal and moisture problems that over time can impact the durability of the building fabric. The first step in improving the sustainability of an older or heritage building is to assess the existing performance: energy, building fabric, and any detailed knowledge of the maintenance and operation of the building. Known deficiencies in historic buildings, such as unwanted ventilation, cold surfaces, and uneven heating, can pinpoint areas requiring attention. Any behavioural issues can be identified and changes made to the use of the building that may illuminate poor practices. Once this has been undertaken, through measurements, analysis of energy bills, and discussions with staff and occupants, the true picture of the building's strengths and weaknesses can be assessed. A plan of interventions that is appropriate for that building can then be assembled and undertaken at a suitable pace.

As mentioned before, fabric upgrades for buildings of this type can be prohibited or unwise, and English Heritage supports the policy of minimum intervention (English Heritage 2012):

The principle of minimum intervention applies at all scales, from an individual brick to works of significant alteration. If all works are kept to the minimum necessary, the maximum historic fabric will be preserved, and thus the significance which it embodies.

The difficulty lies in marrying this long life with current operating conditions and to allow the building to continue to be valued by the owners, the community, and the occupants. Sometimes improving the maintenance of existing structures can offer the chance to improve the performance of a building as a whole. A tabled series of smaller upgrades can lead to continual improvements, small, but each step incremental. Where possible this is the obvious route to take as the reduction in energy use, improvement of thermal comfort, and worth of the building are enhanced for all. Sustainable technologies and innovations (STIs) offer the chance to offset the extra energy used within a historic building with a contribution from renewable sources. Within the debate concerning which STIs to implement, some that would be termed acceptable for newer buildings would be automatically placed in a category of 'difficult to implement' when associated with historic buildings. Less intrusive STIs, both visually and in relation to noise and vibration, are the most favoured for use with historic buildings. Where a low-efficiency heating device is installed, upgrading the heat source alongside pipework insulation and improved control systems can make big differences to emissions and fuel costs alongside maintenance of interior comfort conditions. Solar panels installed on a historic property in a location that cannot be seen from the ground will generally meet the stipulations of local and national regulations and guidelines. The difficulty comes when judging more sensitive cases, where proposed installations might be perceived as intrusive. An array of PV panels sited behind a parapet might be visually acceptable, whereas the parapet might then introduce shadowing for some parts of the day and significantly reduce the effectiveness of the array. Some extensions to historic buildings might, however, be appropriate sites for STIs that in turn reduce the impact and improve the sustainability of the building as a whole.

3.7 Want to know more? Suggested reading and case study sources for upgrading historic buildings

The US National Park Service's Technical Preservation Services department has a series of case studies based upon the use of PV arrays and green roofs on historic properties (US Department of the Interior 2014). English Heritage has a website dedicated to climate change and historic homes, with guidance concerning appropriate upgrades (English Heritage 2011). The UK's Society for the Protection of Ancient Buildings (SPAB) has also published a series of reports that focus on the thermal efficiency of the fabric of historic buildings, building performance, and the very important and linked issue of hygrothermal performance of older buildings (SPAB 2009). A text that focuses upon the choices that architects take when faced with renovating, rejuvenating, or refurbishing historic buildings is *Sustainable Preservations* by Jean Carroon (2011). This text has a wealth of case studies, many of which are very useful, others quite inspirational.

3.4.9   Controls

Effectively controlling the internal environment and any associated technologies and appliances can improve the comfort of building occupants, reduce carbon emissions, and save resources and money for the building owner/operator. Often control systems are overlooked when the promise of modern technologies and innovation offers a forward-looking solution to energy and comfort. The effectiveness of good control systems for heating, hot water, mechanical ventilation, and window opening, shading, lighting, and cooling should not be underestimated.

Most forms of control are systems in their simplest form. A control system consists of three parts: (1) the sensor, (2) the control device, and (3) the environmental technological device to be controlled. The sensor, possibly sensing temperature, illumination or light level, movement, humidity, or air movement, transmits any variance to the controller. The controller takes this input data and transmits a controlling signal to the relevant device or technology. This then varies the output, position, or luminance that impacts the building environment and occupants. The sensors can act in three different ways, through time control, occupancy control, and through controlling condition states. Controlling through time can be very simple, such as through the use of on/off timers that act over one day. More complex systems can run over months with relatively complex settings taking into account vacations and other non-standard states. These often work well when the building has a relatively repetitive schedule such as in schools, regularly opening commercial premises, or indeed the home of a working family. A simple upgrade to improve the energy efficiency, cost effectiveness, comfort, and sustainability of an existing building is to allow new more flexible time controls to enable heating, lighting, and other appliances to be controlled more effectively in relation to true occupation patterns. Where spaces or complete buildings are inhabited intermittently, occupancy control can automatically adjust settings or even switch off some building services entirely. Occupancy controls have often been used to control building services such as lighting and individual ventilation fans that are fast reacting. This aspect of occupancy controls, fast reaction, doesn't currently make them suitable for the control of buildingwide slow-response heating or cooling devices. Probably the most impactful of control systems are those that control by condition, such as those that sense internal and sometimes external environmental conditions. It is most common to bring timers and condition control systems together, such as a domestic building activating a heating system at a particular time in the morning but relying on the temperature control to sense whether the interior temperature is cold enough to warrant boiler activation. Adding occupancy controls can then introduce zone control where rooms or complete floors of a building can have its lights dimmed or even switched off. In larger buildings, with a wide range of building services, such as lifts (elevators), fire/security monitoring, and cooling systems, alongside the more commonly encountered heating and lighting systems, the services can interact. It is then an advantage if these many controls are integrated together in a building energy management system (BMS or BEMS).

rainwater might be considered of high enough quality to drink, this must not be assumed. Rainwater held in appropriate storage can usually be made wholesome by using an integrated system that combines the processes of settlement, filtration, and disinfection (by chemical agents or ultraviolet exposure). Greywater is never fit to drink, but filtration and disinfection can make the water relatively safe to use for limited purposes such as the flushing of toilets.

Rainwater systems are not new. Historically, where boreholes, wells, leats, and watercourses were not sufficient to supply water of either 'quantity or quality', then tanks, barrels, or other storage media could be used to store rainwater. This use could vary depending upon the quality and quantity of water stored. In the past, in isolated farms in areas with clean air, it might have been possible to drink the water, but this is very unlikely today. Large industrial processes that have a requirement for water free from impurities associated with dissolved minerals or other issues benefit from water derived from precipitation. The modern use of rainwater systems has focused upon their use in domestic situations (Figure 3.19). A British standard, BS 8515:2009+A1:2013 (BSI 2009), outlines the major components that normally make up a domestic rainwater system, the recommended quality standards and operating procedures.

Greywater systems differ from rainwater systems in that the water stored or harvested comes from the wastewater. Wastewater can be classified as grey or (foul) black, where the levels of contamination of greywater are substantially less than black water. Greywater commonly comes from the bathing and washing of people and clothes. Foul or black water comes from deeply soiled washing and foul waste from toilets. Greywater systems store and undertake crude treatment of the water, including filtering, settlement, and sometimes chemical or UV treatments. Even treated greywater tends to be cloudy, and the normal usage is for flush toilets. However, as approximately 35% of the potable water used in domestic buildings can be attributed to flushing toilets, this can represent a substantial saving in water. Greywater systems do pose a safety risk, especially if greywater did accidentally become connected to the drinking water system. Very clear connection protocols are needed, along with great care in the training and supervision of operatives installing these systems. Piping that is different in diameter, connection type, and possibly colours can help in the process of keeping any different types of water systems separate.

3.5.4 Sustainable drainage systems

Sustainable drainage systems (SuDs) are an important part of a holistic strategy for the sustainable use and disposal of water from buildings and structures. The main purpose of a SuDS system is 'to minimise the impacts from the development on the quantity and quality of the runoff, and maximise amenity and biodiversity opportunities' (CIRIA 2007/2011). *SuD* normally refers to a system as opposed to one single piece of technology

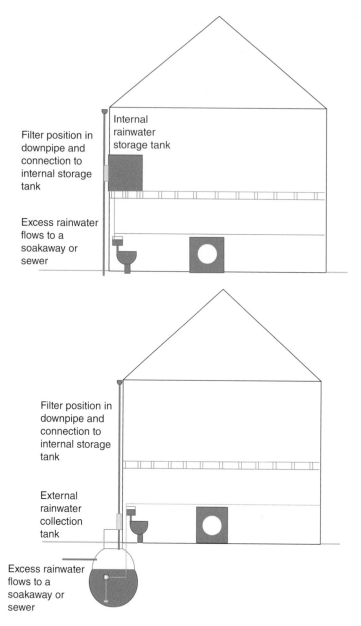

Filter position in downpipe and connection to internal storage tank

Internal rainwater storage tank

Excess rainwater flows to a soakaway or sewer

Filter position in downpipe and connection to internal storage tank

External rainwater collection tank

Excess rainwater flows to a soakaway or sewer

Figure 3.19 Simplified diagram of two rainwater systems with internal and external storage systems.

that will make a traditional drainage system sustainable. This is because the disposal of wastewater has to cope with a number of different sources and categories of water. As has been described within the section focused on grey and rainwater systems, water varies in quality from potable, of drinking quality, to foul water, water heavily soiled, as from toilets. Each of these qualities of water needs to be dealt with in a different way. All SuD systems need to satisfy the requirements of UK legislation, including the 2003 Water Act. This is particularly important if the SuD deals with foul water. The water that needs to be disposed of also comes from different locations in a building's structure and surroundings. A number of the types of technologies that make up a SuD system are described in Table 3.22.

3.5.5 Green roofs and water

Some designers try to minimise a building's physical impact upon the landscape by hiding the space used by people under the ground (earth berming), allowing the ground to be wrapped around and over the building. This is an alternative to conventional construction when the location permits. However, in urban areas this is often not possible. Green roofs are an alternative, and these can offer some of the biodiversity that can be found on earth-bermed buildings. Green roofs and the associated technology of brown roofs are intended to use a flat or nearly flat roof for the introduction of a planted earth area. Brown roofs contain rock or rubble or other materials. The major advantages are shown in Table 3.23.

Table 3.22 The different aspects of a sustainable drainage system (CIRIA 2007/2011)

SuD Technology	Accorded Benefits
Green roofs	Able to absorb and slowly release rainwater. Claimed benefits include an increased water quality
Reed beds	A pond with reeds can act as a form of filter for soiled water.
Ditches, swales (grass-covered shallow channels), and ponds	Often dug surrounding a building or structure, allowing water to be absorbed or held through a tortuous path
Constructed wetlands	Often extended ponds that are normally constantly wet that give wildlife a year-round habitat
Filter strips and drains	Gravel-filled channels sometimes close to a paved area or further away
Pervious paving	Loosely interlocked paving that allows rainwater to infiltrate to the areas under the paving
Soakaways	More traditional method of disposing of waste water

Table 3.23 Major advantages and disadvantages of green roofs

Advantage	Disadvantage
Increased biodiversity	Flat or semi-flat roof necessary
Changed insulation/heat storage properties of the roof area of the building	Increased loadings and therefore possible costs/complexity of superstructure
Reduced rate of run off from green roofs, reducing impact on drainage systems during storms	Possible maintenance issues
Possible increased leisure use of some parts of the roof area depending upon the layout, substructure, and planting system	Planning issues may need to be overcome

3.6   Design for sustainable refurbishment

One of the first decisions confronting a design team when faced with a request from a client to upgrade a facility is whether to demolish or to refurbish. This question becomes more complex when the decision to refurbish is made, as the level of refurbishment needs to be determined. Retaining a building and upgrading it for a similar use or reusing it to perform a different purpose has a number of advantages and drawbacks, as can be seen in Table 3.24.

However, even though Table 3.24 shows that refurbished buildings and structures impose design restrictions, much of the various aspects of the sustainable design of new buildings can be part of the refurbishment of an existing building. These aspects have a different focus when dealing with parts of a building that have been providing the occupants with shelter and comfort for many years. New buildings can be designed to take into account future possible refurbishment. Refurbishment of different classifications of buildings, large, small, bespoke, multiple units all have some common elements that can be taken into consideration for future alteration.

3.6.1   Sustainable refurbishment of domestic buildings

Whilst the number of new homes built in any given year varies alongside the variations of economic and social activity, the number of new housing units very rarely reaches more than 1.5% of all homes on an annual basis

Table 3.24 Advantages of refurbishment versus new build

Advantage of Refurbishing an Old Building or Structure	Advantage of Choosing a New Building or Structure
Keeping a building that could be seen as an accepted part of the landscape	Many of the issues that may have become pressing over the lifetime of the original building or structure, such as public transport routes, could be addressed with a new footprint or innovative design.
If the purpose of the building is largely unchanged, then the social and economic roots that support the use of the building are likely to be left intact.	Old uses for buildings or structures may mean that retaining much of the previous may restrict the possibilities of the new.
Much of the embodied energy that had previously been expended in the materials and construction of the original are intact.	The restrictions of current legislation, particularly thermal regulations, mean it is much easier for a new building to not only to satisfy but also drastically reduce greenhouse emissions compared to a refurbished building.
Historic listings and conservation issues will largely be taken care of, and the ability to show off an older structure as part of an innovative refurbishment can often give a new dimension to the final product that a new build will find difficult to compete with.	The existing structure may not be able to take the weight of taller buildings and therefore restrict the density of the refurbished development when compared to a new structure.

UK Government (2000) *Quality and Choice: A Decent Home for All—The Housing Green Paper*, [Online], Available: https://www.google.co.uk/url?sa=t&rct=j&q=&esrc=s&source=web&cd=1&cad=rja&uact=8&ved=0ahUKEwi00J-72MfJAhXBP RQKHWYKClwQFggdMAA&url=http%3A%2F%2Fwebarchive.nationalarchives. gov.uk%2F20120919132719%2Fhttp%3A%2Fwww.communities.gov.uk%2F documents%2Fhousing%2Fpdf%2F138019.pdf&usg=AFQjCNEXHjFAHoxUiD_ AmsNDm0NFjKnMEQ [6 Dec 2015].

United House (2014) Camden Chalcots Estate PFI regeneration/low-carbon refurbishment project [Online], Available: http://www.unitedhouse.net/projects/id/1275 498417 [24 Apr 2014].

US Department of Energy (2011) *Comparison of Fuel Cells Technology*, Office of Energy Efficiency and Renewable Energy, Washington, DC, [Online], Available: http://www1.eere.energy.gov/hydrogenandfuelcells/fuelcells/pdfs/fc_comparison_ chart.pdf [3 Jan 2014].

US Department of Energy (2013) *2013 Fuel Cell Technologies Market Report*, [Online], Available: https://www.google.co.uk/url?sa=t&rct=j&q=&esrc=s&source= web&cd=1&cad=rja&uact=8&ved=0ahUKEwic27Cu1sfJAhWD7RQKHWl0CHo QFggcMAA&url=http%3A%2F%2Fenergy.gov%2Feere%2Ffuelcells%2Fmarket-analysis-reports&usg=AFQjCNEjbhB-63TEkJPUi3866AI6eJNkbA [10 June 2014]

US Department of the Interior (2014) New technology and historic properties, [Online], Available: http://www.nps.gov/tps/sustainability/new-technology.htm [23 Apr 2014].

Vale, B., & Vale, R. (1975) *The Autonomous House: Design and Planning for Self Sufficiency*, Thames & Hudson, London.

Vale, B., & Vale, R. (2000) *The New Autonomous House: Design and Planning for Sustainability*, Thames & Hudson, London.

Waterwise (2012) *Water Fact Sheet 2012* Waterwise, London. [Online], Available: http://www.waterwise.org.uk/data/resources/25/Water_factsheet_2012.pdf [24 Apr 2014].

Xing, Y., Hewitt, N., & Griffiths, P. (2011) Zero carbon buildings refurbishment: A Hierarchical pathway Renewable and Sustainable. *Energy Reviews* **15**, 3229–3236.

Xu, P., & Chan, E. (2013) ANP model for sustainable building energy efficiency retrofit (BEER) using energy performance contracting (EPC) for hotel buildings in China, *Habitat International* **37**, 104–112.

Zero Bills Homes Company (2013) Introducing the Zero Bills home, [Online], Available: http://www.zerobillshome.com/zerobillshome/ [19 Dec 2013].

4. Materials and sustainable building design

4.1   Materials and design

Materials and their use can have one of the most profound influences upon the sustainability of a building. This influence can be related to the day-to-day and longer-term use of the building, the embodied and the legacy-related issues. The choice of materials can influence the building's thermal performance and internal comfort, acoustics, and durability. The influence is also broad in scope and includes the origins of the material, its production, transportation, longevity, and influence upon the building occupants. The selection of materials can also have economic and social influences on the immediate area around a building project. Within this section the relationship between the choice of materials and building design is discussed. Issues connected with the use of materials and the construction process are dealt with in Chapter 5.

Buildings are, in great part, shaped by the palette of materials selected by a designer to construct a building, as well as the properties that those materials give to the building and surroundings. Materials are used to fulfill the many functional requirements associated with buildings and also to make statements about the building's function, origin, and future. The choice of traditional materials selected, over more modern materials used in the same way, may produce the same basic design but will make a different statement. One common theme that unites most good sustainable building designs is that of the right combination of materials detailed in an effective fashion, suiting the building, occupants, client, climate, site, and surrounding community.

It would be helpful to scrutinise what we expect from a material before discussing which materials would be a sustainable choice and why. Construction materials need to fulfill the functional requirements of a building, mostly based on the building regulations that apply (Table 4.1). Then, to qualify for use in a sustainable building, the material must satisfy a number of further requirements.

Sustainable Construction Processes: A Resource Text, First Edition. Steve Goodhew.
© 2016 John Wiley & Sons, Ltd. Published 2016 by John Wiley & Sons, Ltd.

Table 4.1 Some of the functional and sustainability-related requirements of materials, including the stipulations of the UK building regulations

Traditional Functional Requirements (alongside relevant sections of UK-based building regulations)	Sustainability-Related Requirements
Structural requirements: Part A	Life cycle assessment including quantities of embodied energy
Fire: Part B	Low maintenance/high durability
Sound insulation: Part E	Low waste from construction or little or no cutting/shaping
Good thermal insulative qualities: Part L	Low amounts of harmful pollution from the processes and transport connected with the material (as opposed to purely greenhouse-related emissions)
Affordable for the client/contractor	Skills needed to fit/install available locally
Quality high enough to fulfill client's aspirations	Aesthetic compatible with use
Long/extended life; low carbon emissions	Ability to recycle/reuse

The combination of the requirements shown in Table 4.1 would be challenging for any one material to satisfy. Fortunately, most construction materials and components only have to satisfy some of the traditional functional requirements, due to their relatively specialised role. However, the sustainability-related prerequisites all need to be addressed in some shape or form. There are a number of holistic methods of assessing these requirements and these include whole-building assessment systems, such as BREEAM or LEED and systems approaches. The Natural Step is a systems approach that is a scientific methodology for assessing such diverse things as the impact of changes to practices of legislation. These methods will be discussed in Chapter 6 in the construction assessment section. Of the materials-related assessment systems, life cycle assessment, whole-life carbon, and embodied energy measurements all have a reflection upon the choice of materials for use in sustainable buildings. It is assumed that the mainstream functional requirements of the possible choice of materials have been used to narrow down the choice of construction materials/components so that the more sustainability-related decisions can then be taken.

4.2   Responsible sourcing of materials

Increasingly, the issues related to the selection and procurement of materials have greater scope than life cycle or environmental matters. The sustainable building needs to be not only superior in performance terms but also sound from an ethical, social, and economic point of view. One of the challenges is to demonstrate to all interested parties that the products and materials contained in any construction phase have been sourced responsibly. Judgments on whether a material or product is, in relative or absolute terms, 'good or bad' in relation to its source is not an easy task. From one point of view, these evaluations are linked to the materials used to construct a building, but in many ways they need to be linked to the behaviour of the person who specifies that material. Therefore, the responsible sourcing of materials is also discussed in relation to behaviour in Chapter 7.

The BES 6001 responsible sourcing standard for construction products was developed by the BRE Group in partnership with industry stakeholders. It provides a holistic approach to managing a product from the point at which component materials are sourced through manufacture, processing, and delivery to customers. According to the British Standards Institute, 'The standard describes a framework for the organisational governance, supply chain management, and environmental and social aspects that must be addressed in order to ensure the responsible sourcing of construction products' (BSI 2013). The structure and implications of BES 6001 are described in more detail in Chapter 6. Some texts that include many of the underlying precepts contained in BES 6001 and offer material selection guidance specific to the production and design of buildings are the *Green Building Handbook*, the *Handbook for Sustainable Building*, and the online BRE Green Guide to Specification (Anink *et al.* 1996; Woolley *et al.* 2000; BRE 2014).

4.3   Life cycle assessment

Life cycle assessment (LCA) is a method of measuring the environmental impacts that derive from a material, product, process, or activity. There are standards and guidance, such as ISO 14040/44 (ISO 2006), that define the scope and mechanics of LCA, and the particulars of this are described in more detail in Chapter 6. Most LCA processes consider the potential environmental impacts associated with all the aspects of using a material. There are some accepted aspects of analysis for LCA, such as when during a material's life the analysis should start and when it should cease (if it ceases). For example, the term *cradle to grave* refers to the assessment of environmental impacts from the extraction of the raw materials that are used in the material's manufacture through its use and finally to its disposal. Other time frames can include a later final point if it is assumed that the material or product becomes reused or recycled: cradle to cradle, cradle to gate, well to wheel (Table 4.2).

Following agreed time frames and a standardised methodology, possibly using a piece of commercially available or bespoke software, a building professional can establish the environmental impacts of different materials and products. As with most standardised processes, there are, however, some points in time when the real world has to be approximated so that the standardised assessment can work. Decisions are made about the boundaries of the production impact, transportation, use, and disposal of a material so that an impact can be attributed to that material. One method of exploring the boundary is to take the example of a building, its production, use, and disposal. In the case illustrated in Figure 4.1, the system boundary has been drawn in a specific place that includes specific activities and flows intentionally associated with the building, but it cuts out any

Table 4.2 Showing the range of life cycle assessment (LCA) categories

Form of LCA	Assumed Boundaries and LCA Characteristics
Cradle to grave	LCA includes all inputs and outputs surrounding the obtaining of, the use of, and the disposal of the material or product
Cradle to gate	As above, but the LCA is of part of the material or product's life cycle, evaluating only the material or product life cycle that takes the assessment to the gate of a factory. The LCA elements associated with use and disposal are not included. Whilst this might seem a retrograde step, it does allow separate LCAs for separate materials and products to be accumulated and then used by others.
Gate to gate	Similar to cradle to gate, this is a further reduction of the scope of the LCA, and meant to enable comparisons of one point of production or processing with another. In this approach only raw materials and production/processing energy alongside emissions and waste attributed to one facility are included.
Cradle to cradle	This form of LCA is often associated with a closed-loop approach, where the final part of the material or products life is for it to be recycled (assuming that if it could be reused that this has already happened).
Well to wheel	Rather a specialised form of LCA but quoted relatively frequently; this refers to fuel and in the main vehicles. However, as some standalone generation may well use a range of delivered fuels including biofuels, it is felt that this form of LCA should be included in this table. Whilst currently this system is used to estimate the impacts of the choice of different fuels used when transporting items, similar choices could be routine for the designers and managers of buildings. Well to wheel takes the journey from the origin of the fuel, a 'well' or feedstock, through all the processes needed to then allow the transportation in question to function. Similar subdivisions of cradle to cradle can be applied, splitting the journey of the fuel and its combustion, such as 'plug to wheel', that takes the efficiencies of the combustion cycle of a typical engine through the various losses in transmitting the power to the wheel of a vehicle.

activities or flows that could be categorised as less pertinent or unrelated to the building. This is a difficult task, and more discussion and descriptions of the standards that grapple with this issue are to be found in Chapter 6.

As an LCA example, consider a building designer choosing between timber or steel for a structural frame. Initially, if the designer were concerned about using a more natural material (assuming natural materials were automatically more sustainable), timber might be taken as first choice, but on closer inspection the decision is quite complex. The material's properties need to be taken into account, as listed below.

Mass: The mass of materials needed to undertake a particular role in a building can vary considerably, and a steel/concrete/timber structural member might vary, leading to surprising results if the overall embodied energy of a component is compared.

Lifetime: Product A lasts X years whilst Product B lasts X + 20 years.

Site-based inputs: If a material has been selected on the basis of cradle to gate or even on a cradle-to-site basis, if the material requires further energy input in the form of machining to ensure fit, or site-applied coatings for durability of fire resistance, this may alter any selection decisions.

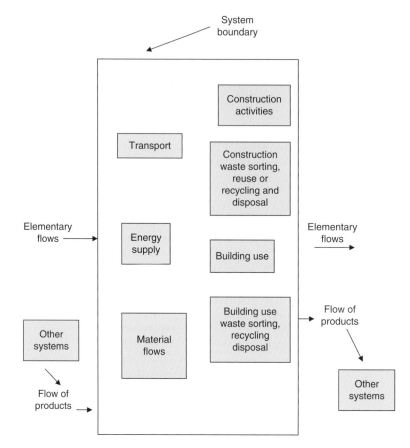

Figure 4.1 A system boundary for using life cycle assessment associated with a specific building or project.

Maintenance: Will the selected material need further energy/material/people-related inputs to ensure that this material will keep functioning at the assumed level over its lifetime?

Waste: If 1 kg of material is used in the building, but 20% or 200 g is wasted due to cutting and shaping, how does this change the decision to select or not select? This is further complicated through the disposal/recycling/reuse options for the waste.

4.4 Whole-life carbon (embodied energy and embodied carbon)

To allow many countries, including the UK, to meet an objective of reducing their carbon dioxide emissions by at least 80% by 2050, against a 1990 baseline, and with significant progress to be made by 2020, the construction industry will have an important role to play. Nondomestic or commercial buildings are responsible for 18% of the UK total carbon emissions (Carbon Trust 2009). If this is added to emissions related to domestic buildings, the impact of reducing these is considerable in terms of operational energy.

4.1 Want to know more? Embodied carbon

One publication that is very helpful when trying to calculate these embodied quantities is the BSRIA guide, Embodied Carbon (Hammond & Jones 2011). The BSRIA guide is based on the ICE open-source database, which is an ideal resource for any carbon design tool. To use the most up-to-date data, readers will need to refer to the latest version of the ICE database at www.bath.ac.uk/mech-eng/sert/embodied (University of Bath 2014). The BSRIA guide is based on version 2.0 of the ICE database. The ICE database will be mentioned again later in this section.

As the drive for more energy-efficient buildings, both commercial and domestic, gathers momentum through the increasing number of better performing new-build project completions and refurbishments, the quantity of energy associated with the operation of buildings is predicted to fall. The emissions that can be ascribed to the energy needed to produce materials, and operations to construct and demolish these buildings, will become more significant in comparison with operational energy.

Several recent texts, initiatives, governmental groups, professional bodies, and pieces of academic research now guide any interested parties to important facts, methods, and actions that will allow the measurement and targeting of embodied energy and carbon. The measurement of energy use in a building is not an easy task, when the major use of the energy needs to be identified under headings such as appliances, people, services, and fabric. However, this is really quite straightforward when compared with the task of assessing the non-operational energy attributed to a whole building. These difficulties are eased by using an agreed standardised methodology. BS EN 15978:2011 (BSI 2011) was developed by the European Committee for Standardisation (CEN) Technical Committee 350 (TC350). This is a good reference point and has been used by a number of other pieces of guidance related to whole-life carbon. The first task is to define what is meant by embodied energy and carbon.

The BSRIA Guide defines embodied energy (carbon) as:

> The total primary energy consumed (carbon released) from direct and indirect processes associated with a product, or service, and within the boundaries of cradle-to-gate. This includes all activities from material extraction (quarrying/mining), manufacturing, transportation, right through to fabrication processes, until the product is ready to leave the final factory gate.
>
> (Hammond & Jones 2011)

As described above, in LCA the agreed energy boundaries have a large influence upon the final figure, whether energy or carbon. But if those boundaries are agreed, whether cradle to grave, or another start and finish point, the issue of what constitutes energy and what constitutes carbon (emissions) also has to be established. Once this has been quantified, each part of the embodied emissions may be equated to an amount of carbon dioxide, and the cumulative total becomes the whole-life carbon embodied in the building. As the amount of energy embodied in a building, which is not easily determined, is normally less debatable than the quantity of emissions ascribed to that energy use, this text will start the processing of quantifying embodied carbon by examining embodied energy and move on to whole-life carbon.

4.4.1   Embodied energy

Embodied energy is the amount of energy, measured in MJ/kg, that is used in the processing and transport of a material from its raw state to a finished product used within or as part of a building. The boundaries used to assess what activities and their related energy should be contained in the final product are important. Most systems of assessing embodied energy rely on beginning and ending the energy 'count' at the 'birth' or 'death' of a product. The commonly used terms associated with these boundaries are described in Table 4.2.

An associated measurement of materials is embodied carbon ($kgCO_2$/kg). This is the amount of carbon (or carbon dioxide equivalent) released into the atmosphere as a result of processes related to the production and placement of material-related products (Alcorn 2003).

When comparing the embodied energy attributed to different materials or products, there are stated methodologies that should be used when calculating these energy quantities. These methodologies allow one assessment of embodied energy to be compared with another. However, an assessment of materials that could be used in a building also encompasses the potential greenhouse impacts. A number of tools exist that use the Greenhouses Gas Protocol's corporate standard to obtain a fair evaluation of different materials (Greenhouse Gas Protocol 2012). The most comprehensive listing of embodied energy for construction materials commonly used in the UK has been undertaken by the University of Bath under the Carbon Visions Building Programme and is the foundation for the previously described BSRIA Guide, *Embodied Carbon: The Inventory of Carbon and Energy (ICE)* (Hammond & Jones 2011).

This document is an inventory of embodied energy and carbon coefficients for building materials and uses a database of energy and carbon-related figures. This database contains over 1700 records using data from publicly accessible sources, such as journal and conference proceedings. The authors state that the database has a greater confidence in the accuracy of embodied energy figures than that of embodied carbon figures. The rationale for this difference in confidence arises because different fuels can be used to produce a given building material. Whilst a construction material could (within reason) have its embodied energy quantified with some certainty, factories using fuels (feedstock energy) such as coal, natural gas, hydroelectricity, or some other power source will have differing carbon emissions related to the same embodied energy. Whilst this might introduce an element of doubt in relation to the final embodied energy or carbon figures associated with building materials, the ICE database is open in relation to the methodology used and sources for the data contained within it. The authors have also endeavored to use data that was compliant with ISO 14040/44 (ISO 2006), the International Standard on Environmental Life Assessment, and use a system boundary based upon a cradle-to-gate methodology. The number of cited users is also large, but the quality of the citations tends to indicate that the ICE database is one of the best sources currently available.

A few examples of the relative values for some of these materials can be seen in Table 4.3.

Table 4.3 has three columns showing three distinct but linked numbers related to the amount of energy and the fuel used in production. They are Embodied Energy (MJ/kg), Embodied Carbon ($kgCO_2$/kg), and Equivalent Embodied Carbon ($kgCO_2$e/kg). The three columns give data that might be used to build up the quantity of greenhouse-related emissions from either the embodied energy (Column 1) or the CO_2 emissions (Column 2). An alternative is to use the CO_2e figure in Column 3. This uses a cumulative figure ascribing the greenhouse impacts from all the emissions (both non-carbon dioxide, such as methane, CFCs, SO_x and NO_x, along with the often larger quantities of carbon dioxide gas) either to one figure or two, the first ($X.XX_{fos}$) to fossil fuels and the second ($X.XX_{bio}$) to biofuels (where X refers to the figure ascribed to the appropriate fossil or bio fuel). Before using these figures to calculate the embodied energy that might be associated with some typical building elements, it is logical to first explain how using embodied carbon and whole-life carbon can be a next step in determining the embodied energy of a material, component, building element, and finally a building. Let us now look in more detail at the embodied energy within a typical masonry wall specification that might easily be encountered in the UK and compare this with a straw bale wall.

4.4.1.1 Examples of embodied energy and related calculations

Recent publications provide information for the calculation of embodied energy/carbon for a part or a whole building. As previously mentioned, the ICE database (Hammond & Jones 2011) is a good source of data in relation to individual materials. These can be used to contribute to whole-building calculations or to explore the relative merits of different configurations of materials that might be used for the same building element. The final report of the UK government's BIS Low Carbon Construction Innovation & Growth Team (IGT) (HM Government 2010) gives some good guidance concerning strategic directions. This advocates that whole-life carbon could become an equivalent measure, equal to whole-life cost. A Royal Institute of Chartered Surveyors (RICS) information paper that focuses upon the methodologies for calculating the embodied carbon as part of the life cycle carbon emissions for a building refers to the IGT documentation. This information paper aims to offer practical guidance to quantity surveyors and is focused upon calculating cradle-to-gate embodied carbon emissions associated with building projects in the UK (RICS 2012).

Table 4.3 The embodied energy for some commonly used construction materials (Hammond & Jones 2011). Reproduced by permission of BSRIA

Material or Group of Materials	Embodied Energy (MJ/kg)	Embodied Carbon (kgCO$_2$/kg)	Equivalent Embodied Carbon (kgCO$_2$e/kg)
Aluminum (extruded)	154	8.16	9.08
Brick (common)	3.00	0.23	0.24
Cement (UK weighted average)	4.5	0.73	0.74
Mortar (1:1:6 cement:lime:sand)	1.11	0.163	0.174
Mortar (1:4)	1.11	0.171	0.182
Mortar (1:5)	0.97	0.146	0.156
Mortar (1:6)	0.85	0.127	0.136
Cement stabilised soil @ 5%	0.68	0.060	0.061
Concrete	0.75	0.100	0.107
Reinforcement for concrete	1.04	0.072	0.077
Cellular glass	27.00	—	—
Cellulose	0.94 to 3.3	—	—
Cork	4.00	0.19	—
Fibreglass (glasswool)	28.00	1.35	—
Flax (insulation)	39.50	1.70	—
Mineral wool	16.60	1.20	1.28
Paper wool	20.17	0.63	—
General	5.30	0.76	0.78
Damp-proof course/ membrane	134	4.2	—
Paint double coat	21.0 MJ/Sqm	0.73 kgCO$_2$/Sqm	0.87
Plasterboard	6.75	0.38	0.39
Polyurethane rigid foam	101.50	3.48	4.26
Sand (fine aggregate)	0.081	0.0048	0.0051
General (rammed soil)	0.45	0.023	0.024
Steel—UK (EU) average recycled content	20.10	1.37	1.46
Granite	11.00	0.64	0.70
Timber (general value)	10.00	$0.30_{fos} + 0.41_{bio}$	$0.31_{fos} + 0.41_{bio}$
Glue laminated timber	12.00	$0.39_{fos} + 0.45_{bio}$	$0.42_{fos} + 0.45_{bio}$
Plywood	15.00	$0.42_{fos} + 0.65_{bio}$	$0.45_{fos} + 0.65_{bio}$
Sawn hardwood	10.40	$0.23_{fos} + 0.63_{bio}$	$0.24_{fos} + 0.63_{bio}$
Medium density fibreboard	11	$0.37_{fos} + 0.35_{bio}$	$0.39_{fos} + 0.35_{bio}$
PV module monocrystalline	4750 (2590 to 8640)	242 (132 to 440)	—
PV module polycrystalline	4070 (1945 to 5660)	208 (99 to 289)	—

Note: Dashes denote information that is not available.

As shown in Table 4.10, the embodied energy attributed to the traditional masonry wall is approximately three times that of the straw bale wall. The thermal transmissions through the two walls are much closer in value, with the straw bale wall being slightly better, allowing fewer Watts per square metre per degree Kelvin to be conducted through it. However, the choice of wall

4.2 Want to know more? Comparing embodied energy

To provide the reader with a comparison of the embodied energy contained in two different types of walling, the next calculations show the methodology for comparison of different options for a unit area of traditional domestic wall compared with a straw bale wall. The calculation is presented as a simple demonstration of the theoretical aspects behind embodied energy and carbon rather than closely following an ISO standard version of this form of calculation. Table 4.4 lists the materials commonly found in a masonry cavity wall and their characteristics, using assumptions shown in Table 4.5, in part according to the University of Bath's figures (Hammond & Jones 2011). Tables 4.6 and 4.7 work in very much the same way but for a straw bale wall.

In summary, the traditional masonry insulated cavity wall has 1525 MJ/m² energy embodied in each square metre compared with the result for a straw bale wall of 580 MJ/m²: a considerable difference. If a more holistic view of the comparison is introduced, where the thermal transmission value (U-value) is calculated, the potential occupational emissions can be compared with the embodied emissions. Again, this calculation is undertaken to compare the thermal properties of one walling system with another rather than to calculate a U-value ready to be used in regularity discussions. Therefore, the dimensions are approximate and the elements such as mortar joints and cold bridging are not included to enhance the simplicity of the demonstration.

Table 4.4 The embodied energy of each material used in the unit area of insulated masonry cavity wall (energy and carbon data from Hammond & Jones 2011) *Values from ICE*

Traditional Cavity Wall, Material/Layer	Energy MJ/kg	Carbon kg CO_2/kg	Density kg/m³	Embodied Energy (MJ/kg × kg/m²) MJ/m²
Render	0.85	0.136	1800	31
Block	0.75	0.107	2200	132
Insulation (fibreglass)	28	1.35	100	280
Brick	3	0.24	1900	371
Plasterboard	6.75	0.38	1800	164
Paint MJ/m²	21	0.73		21
Mortar	0.97	0.146	1800	524
PVC DPC	134	4.2	1500	0.33
Stainless steel wall ties	56.7	6.5	3000	2.41
	Total embodied energy (MJ/m²)			1524

specification, much as the choice of the specification of the many other elements that make up any building, often needs to take into account a range of other factors. In this case the wall thickness of the straw bale wall is approximately 470 mm, when the render and rain-screen are taken in to account, whilst the masonry wall is approximately 330 mm. This extra 140 mm thickness, added on all four walls, increases the building footprint (or reduces the internal dimensions of rooms) and might restrict the development of a potential site.

Table 4.5 Showing the assumptions used in the calculation shown in Table 4.4

Assumptions	kg/m²	Thickness (m)
Render	36	0.02
Blocks	176	0.1
Insulation	10	0.1
Bricks	123.5	0.1
Plasterboard	24.3	0.0135
Mortar for blocks	360	1
Mortar for bricks	540	1
Paint for 1 m²	21	
DPC[a]	13.27	0.33
Steel ties[b]	2.41	

[a] Assumes a 6-m-high wall and DPC on inner and outer wall average is 2 m run over 6 m2, so this allows 0.333 m DPC per 1 m2 of wall.
[b] Assumes 1 m run of 3 mm stainless wire ties in each 1 m² section of cavity wall.

Table 4.6 The embodied energy of unit area panel, similar to Table 4.4 but of materials of a more innovative nature, for this example a straw bale wall (energy and carbon data from Hammond & Jones 2011)

Straw Bale Wall, Material/Layer	Energy MJ/kg	Carbon kg CO₂/kg	Density kg/m³	Embodied Energy MJ/kg × kg/m² MJ/m²
General timber rainscreen	10	0.71	900	162
Lime render	5.3	0.76	1800	191
Straw	0.24	0.01	100	9.6
Timber load-bearing frame	10	0.71	300	0.15
Paint MJ/m²	21	0.73		21
Steel timber and straw fixings	20.1	1.37	3000	5.11
PVC DPC	134	4.2	1500	0.33
Total embodied energy (MJ/m²)				580

Handwritten margin notes: Energy + Carbon figures come from ICE. NGV load bearing. Assumed typical. Lime render – for FIRE REASONS. ⅓ Embodied energy for same U Value of Masonry Wall

Straw bale construction on a constricted city site might therefore be less economic. The specification of the masonry wall might be altered so as to bring down the amount of embodied energy but still allow the thermal values and footprint-related issues to be minimised.

4.4.2 Embodied carbon and whole-life carbon

As discussed in the previous section, operational emissions from buildings, due to energy consumption for lighting, heating, ventilation, and air-conditioning, have reduced over recent years through better design and management. The carbon dioxide emissions from the manufacture

Table 4.7 Showing the assumptions used in the calculation shown in Table 4.6

Assumptions kg per m²	kg/m²	Thickness
General timber	16.2	0.018
General timber from the Bath ICE index (mix of typical timbers used in the UK)		
Lime render	36	0.02
Straw	40	0.4
Lime render	36	0.02
Timber load-bearing frame[c]	0.015	
Paint for 1 m²	21	
DPC[a]	13.266	0.333
Steel timber fixings[b]	0.25	kg/m²

Volume of Timber	Upright	Horizontal
Dim1	0.2	0.1
Dim2	0.05	0.05
Height	6	6
Volume over 6 m	0.06	0.03
Total for both	0.09	M³ per 1 m run
Volume per m²		0.015

[a] Assuming a 6-m-high wall and DPC on inner and outer wall average is 2 m run over 6 m2 so this allows 0.333 m DPC per 1 m2 of wall.
[b] Assuming an average of X6 3 mm nails, each 100 mm long in each 1 m² section of cavity wall.
[c] Timber frame assumed to be a mixture of slim 200 mm by 50 mm and 100 mm by 50 mm components to stop the frame interrupting the continuity around the building. Average volume of timber is assumed to be one 200 mm by 50 mm upright and one 100 mm by 5 mm horizontal over a 1-m run of a 6-m-high wall.

Table 4.8 Thermal transmission (U-value) calculation for a masonry cavity wall

Traditional Masonry Wall Material	Thickness m	Thermal conductivity W/m K	Resistance m²K/W
Internal surface resistance	n/a	n/a	0.123
Render	0.02	1	0.025
Block	0.100	0.12	0.8
Insulation (fibreglass)	0.100	0.02	5
Brick	0.100	0.84	0.12
Plasterboard	0.013	0.14	0.093
External surface resistance	n/a	n/a	0.055
Total thermal resistance			6.25
U-value W/m²K			0.16

and installation of building products and materials is now the second most significant area of carbon emissions from a building over its life cycle (HM Government 2010). Understandably, sustainable buildings now need to reduce their tally of embodied carbon. To do this, the original operational energy that was expended in creating the materials and products,

Table 4.9 Thermal transmission (U-value) calculation for a straw bale wall

Straw Bale Wall Material	Thickness m	Thermal conductivity W/m K	Resistance m²K/W
Internal surface resistance	n/a	n/a	0.123
General timber rainscreen	0.018	0.5	0.03
Lime render	0.020	0.24	0.083
Straw	0.400	0.06	6.67
Timber load-bearing frame	n/a	n/a	
Lime render	0.020	0.24	0.083
External surface resistance	n/a	n/a	0.055
Total thermal resistance			7.04
U-value W/m²K			0.14

Table 4.10 The comparison between a traditional masonry wall and a typical straw bale wall, embodied energy and thermal transmission

Wall Construction	Embodied Energy MJ/m²	U-Value W/m²K
Traditional masonry insulated cavity wall	1525	0.16
Straw bale wall	580	0.14

and the energy used in the construction, need to be effectively quantified in the form of tonnes CO_2e. The total quantity of embodied carbon (shorthand for carbon dioxide equivalent) in a building can then be judged against equivalent buildings and then a measure of the impact of the building established. However, according to the *Low Carbon Construction Innovation & Growth Team: Final Report*, this process should have due consideration for possible areas of wasteful practice dressed in the guise of low-carbon materials or processes:

> The one qualification to this. Given that a key objective must be energy reduction, rather than just switching to energy from cleaner sources (which may still be used wastefully) or to carbon trading, then energy should be measured first when assessing the performance of a building, and then be converted to its CO_2 equivalent.
>
> (HM Government 2010)

The need to take into account the level of the embodied carbon in a building is therefore necessary. However, the calculation or estimation of embodied carbon is complex. This complexity, wedded to the fact that this is a relatively new area of research, means that a number of assumptions have to be made that will affect the accuracy of the outcome (much as the same types of assumptions are needed in relation to embodied energy estimations). However, there are some agreed and established frameworks that allow embodied carbon estimations to be comparable with other estimates, assuming that all methodologies are made as transparent as possible. In time, like-for-like comparisons will become more accurate and consensus will be reached.

The first hurdle to comparing the embodied 'level of impact' of one material/process/activity with another is the fact that not all of these will emit carbon dioxide. Some activity may emit other gases linked to a different level of global warning. The ability to link emissions to their impact has been formalised through the use of a concept/system referred to as global warming potential. As carbon dioxide CO_2 is the most prevalent global warming related gas and when compared with most other global warming gases has the least potential for damage, it is allocated a potential of one, and all other gases are measured from that datum. The IPCC definition relating to global warming potential (GWP) states,

> Global Warming Potentials or other emission metrics provide a tool that can be used to implement comprehensive and cost-effective policies (Article 3 of the UNFCCC) in a decentralised manner so that multi-gas emitters (nations, industries) can compose mitigation measures, according to a specified emission constraint, by allowing for substitution between different climate agents.
>
> (IPCC 2007)

Thus, GWPs allow for comparisons between different gases linked to global warming and so give a relative importance to each gas over a number of years. Table 4.11, published by the IPCC, shows the large differences in GWP between different gases and the changes of that potential over three time periods, 50, 100, and 500 years.

Thus, using Table 4.11, if an equal quantity of a commonly encountered refrigerant such as HFC R23 were emitted alongside an equal quantity of carbon dioxide, the impact of the R23 in relation to GWP (at a time span of 50 years) would be 264 times that of the CO_2. The incidence of R23 emissions is very much lower than the incidence of CO_2; thus, the higher potential impact of R23 gas emissions would be, in reality, much less that CO_2 emissions. Once the GWPs of the different gases have been equated to an equivalent tonnage of carbon dioxide emissions (tonnes CO_2e), the process of calculating the relative impacts of different materials/activities/processes can begin.

4.4.2.1 Strategies to reduce embodied energy and carbon from buildings

Many initiatives and tools have been introduced to describe the actions and provide methodologies for reducing the quantities of embodied energy and carbon in our buildings. The UK government's Low Carbon Construction Action Plan aims to improve the government's own performance in this area, eventually influencing the performance of the wider construction industry. The plan comments on the need for government to balance incentives and 'interventions', many elements relating to the need for the Green Construction Board (GCB) to lead the Action Plan in tandem

Table 4.11 Global warming potential referenced to the updated decay response for the Bern carbon cycle model and future CO_2 atmospheric concentrations held constant at current levels

Greenhouse Gas Species	Chemical Formula	Lifetime (Years)	Global Warming Potential (Time Horizon)		
			20 years	100 years	500 years
CO_2	CO_2	variable§	1	1	1
Methane*	CH_4	12 ± 3	56	21	6.5
Nitrous oxide	N_2O	120	280	310	170
HFC-23	CHF3	264	9100	11700	9800
HFC-32	CH2F2	5.6	2100	650	200
HFC-41	CH3F	3.7	490	150	45
HFC-43-10mee	C5H2F10	17.1	3000	1300	400
HFC-125	C2HF5	32.6	4600	2800	920
HFC-134	C2H2F4	10.6	2900	1000	310
HFC-134a	CH2FCF3	14.6	3400	1300	420
HFC-152a	C2H4F2	1.5	460	140	42
HFC-143	C2H3F3	3.8	1000	300	94
HFC-143a	C2H3F3	48.3	5000	3800	1400
HFC-227ea	C3HF7	36.5	4300	2900	950
HFC-236fa	C3H2F6	209	5100	6300	4700
HFC-245ca	C3H3F5	6.6	1800	560	170
Sulphur hexafluoride	SF_6	3200	16300	23900	34900
Perfluoromethane	CF_4	50000	4400	6500	10000
Perfluoroethane	C_2F_6	10000	6200	9200	14000
Perfluoropropane	C_3F_8	2600	4800	7000	10100
Perfluorobutane	C_4F_{10}	2600	4800	7000	10100
Perfluorocyclobutane	c-C_4F_8	3200	6000	8700	12700
Perfluoropentane	C5F12	4100	5100	7500	11000
Perfluorohexane	C6F14	3200	5000	7400	10700

§ Derived from the Bern carbon cycle model.
*The GWP for methane includes indirect effects of tropospheric ozone production and stratospheric water vapour production.
Source: IPCC Fourth Assessment Report: Climate Change 2007 (IPCC 2007).

with other UK construction-related bodies, such as professional bodies, the UK Green Building Council, and the Strategic Forum for Construction. The GCB's intention is to move forward the low carbon agenda in key areas, including the design, development, financing and operation, of low carbon buildings and infrastructure. The GCB will produce a Low Carbon Route map for the built environment, setting out key milestones and interventions for the delivery of an 80% reduction in emissions by 2050. As the GCB is an association of Government and Industry, understandably its goals feed into the UK's need to fulfill national responsibilities both on European and global stages. Some of the board's desired outcomes come from the 'Fundamental Truths' section of the resources on the GCB website (GCB 2012). These include:

- A contribution to climate change mitigation
- Reduced use of resources
- Lower embodied carbon and capital cost
- Lower operational carbon and running costs
- Improved solutions
- Increased UK competitive advantage

Some of these desired outcomes might be considered general in nature but do ally with overarching governmental responsibilities, including allowing the UK economy to grow and increase/maintain construction employment alongside the wider sustainable construction drivers.

4.4.2.2 Examples of whole-life carbon and related calculations

The individual carbon content embodied in building elements follows the logical sequence of calculations described in the earlier section, which focused on establishing embodied energy. The use of the figures in the BSRIA/Bath ICE database (Hammond & Jones 2011) relating to the equivalent embodied carbon, $kgCO_2e/kg$, rather than the energy-related figure, will allow an overall indicator of how much carbon dioxide and other global warming emissions can be pinned upon a building element. However, this can be scaled up to include a whole-life carbon figure for a complete building. The RICS has published a draft information paper that describes a methodology for the calculation of embodied carbon as part of the life cycle carbon emissions for a building (RICS 2012). This paper logically infers that any whole-life carbon estimation needs to follow exiting life cycle guidance and methodologies. Thus, BS EN 15978:2011 (BSI 2011) is felt to be the chosen calculation method used by the construction industry, and therefore the paper adopts the life cycle classification from the BS EN standard.

Through the previously mentioned IGT report (HM Government 2010), the contribution from each of the stages in a building project's carbon life cycle emissions can be attributed as shown in Table 4.12.

These are complimented by some further clarification by the European Committee for Standardisation Technical Committee 350 (BSI 2011). As can be seen from Table 4.11, the majority of the $kgCO_2e/kg$ can be attributed to the manufacturing phase after the operations phase. This ratio will, however, alter depending on the use of the building, the level of input from building services, and the size and type of structure. A largely unheated building built from products and materials that have a very high embodied carbon equivalent might have a ratio that could approach as much as 50% embodied to 50% operational. Whilst there are embodied emissions in phases other than phases other than manufacture and operation, their magnitude is currently relatively small. As previously mentioned, this will change in the future as continuing efforts are made to reduce the influence of manufactured embodied emissions and the regulatory pressure on energy use through wider energy efficiency measures become more common place. The overall emissions

Table 4.12 A building project's carbon life cycle emissions according to process or stage of building production or use

Phase of Construction Project	Estimated Percentage of Carbon Equivalent Embodied at that Phase (% of $kgCO_2e/kg$)
Design	0.5
Manufacture	15
Distribution	1
Construction	1
Operation	83
Refurbishment/demolition	0.4

related to embodied and operational aspects of buildings are likely to reduce, and the eventual goal will be to minimise the embodied emissions whilst also reducing the operational emissions. As part of this process, the calculation of whole-life carbon and the ability to identify this aspect in any construction project and compare the result against best practices will help to further this goal of emissions reduction. From the old stage D (roughly equivalent to stage 3 of the 2013 version) RIBA plan of work, the RICS draft information paper relating to embodied carbon proposes a very rational methodology to encompass the whole-life carbon of a building (RICS 2012).

4.4.3   Low maintenance and high durability

Interior and exterior building construction materials both have durability issues connected to their function and position. Traditionally, exterior materials placed in exposed positions around a building are the most vulnerable to decay and wear. Interior finishes can also have an arduous role to play, particularly floor finishes in buildings like hospitals, where good resistance to dirt to prevent spread of infections is a very necessary requirement. The difficulty when trying to select materials that are both durable and offer low maintenance whilst ensuring a sustainable choice, is that many natural products are not as durable as many manmade products. This statement should, however, have caveats. Some perceived natural products, such as linoleum floors, can be very durable whilst others can be durable if the correct maintenance system is employed. Some vulnerable more sustainable materials, such as fast-grown softwood window frames, will require frequent application of appropriate protective stain or paint to ensure they last and, just as importantly, function throughout their lifetimes. Here good design can reduce possible maintenance, detailing the window and its surrounds to ensure that as little driving rain as possible falls upon the glazed unit and any rain is directed away from the timber sections that make up any framework. This highlights yet another minor conflict: ensuring a building is sustainable in a certain aspect—durable windows—whilst ensuring the building maximizes its internal space. One of the methods of improving the durability of most timber framed windows is to set them back into the wall, reducing potential internal space. This could just be a minor inconvenience for a domestic building, or it could lead, in other instances, to a reduction in the lettable area for a commercial building.

4.4.4 Low waste from construction with little or no cutting and shaping

The low-waste aspect of sustainable construction materials will be discussed in Chapter 5, which focuses upon the construction process, but some decisions made during the design process impact upon this issue. Many materials are produced in small units so they can easily be produced and moved to a construction site; in many instances they are are small enough to be handled and placed by builders. This has led to the use of standard sizes related to particular types of components such as bricks, blocks, and plasterboard. This fact, when combined with the issue that much of the construction waste that flows out of construction sites is related to cutting materials to fit, focuses upon the issue of waste in relation to design. If standard units are available, then designs can use this dimension as a possible aid to ensuring, where possible, that dimensions of all parts of the building are multiples of this dimension. This idea is not new. Modular building systems have been tried before and had some degree of success and failure. What is new is the higher price that wasting materials now entails. The client ends up paying not only for lost materials and the labour to cut/shape the materials to fit but also the high charges that will be applied to the waste that leaves the site destined to landfill or incineration. However, standardised materials are not suitable for every construction project, as some projects involve considerable amounts of bespoke units, often due to the building's prospective use or the demands of a 'difficult' site.

Tolerances, when tackled at the design stage, can also have an impact on waste from materials. Some materials, such as steel framing, have to be placed within a high degree of accuracy. The steel frames are normally fixed to 'holding-down' bolts set into concrete bases, a relatively low accuracy situation. If at the design stage suitable allowances can be built into the process, allowing for the mismatch of these two degrees of accuracy, then wasted material/time, such as welding or cutting out of bolts, is less likely to happen. Using fixing systems that can take into account the likely site conditions and the ability to put right any difficulties will often not only reduce materials wastage but also improve the predictability of the completion stages of any constructed building.

4.4.5 Pollution generated by the manufacture of materials and products

The reduction of waste during the construction of buildings and projects is in many ways within the control of construction professionals. However, the pollution that occurs from a manufacturing process is often removed from the team of people that procure, design, construct, and maintain our buildings (localised pollution that is confined to a site and related to construction processes might well include dust, and this is discussed in Chapter 5, in the section titled 'Dust', including the instance of silica and non-silica dusts). Very similar building materials and products may be produced by a number of manufacturers using different or even similar processes. Of these producers, some may take more care over the sourcing of their materials, the efficiency of their processes, and the reduction in failures due to poor quality control than others. While all manufacturers are guided, or compelled to adhere to legislation and regulations, this adherence can vary. The variance can sometimes be attributed to some extent to outsourcing parts of the process to less expensive subcontractors who may have not only different national regularity frameworks to comply with but also a different view of corporate and social responsibility. Some companies, however, will take all steps necessary not only to comply but also to adopt best practice. This is normally associated with membership or accreditation to one or more of the sustainable building–badged associations that are in turn recognized by national or international schemes. Using the materials and products manufactured by environmentally responsible companies that have been externally accredited will often incur a monetary cost, as the extra measures needed to produce such products result in extra overheads. However, over time, as any perceived or actual risk factors associated with innovative products reduce, other companies working in this field start to adopt better working practices, and market penetration increases for the best materials and products, the costs are likely to reduce.

4.4.5.1 Low amounts of harmful pollution from the processes and transport connected with the material (as opposed to purely greenhouse-related emissions)

Some construction products are produced from raw materials that are relatively benign, to the building occupants when the building is completed, to the operatives who construct the building, and to the people who live near the plants that produce and finally deal with the products after use. However, in many cases this is not true. There are a number of materials that can be harmful in a fairly crude and unspecific way, including PVC (USGBC 2007), materials containing formaldehyde, adhesives containing high concentrations of VOCs, and phenols. Plastics have been and are increasingly becoming part of the spectrum of materials used in the structure, insulation, fixtures, fittings, and finishes of a building. The British Plastics Federation (BPF) has information concerning the recycling of most plastics and has also launched an initiative to 'showcase the UK as global leaders of sustainable manufacturing in plastics processing'. Whilst the overt goal is to portray the industry in a good light, there are some valuable facts and figures in the guides on this organisation's website (BPF 2014a; BPF 2014b). In contrast to the pollution that some construction materials cause when they are produced or disposed of, other health-related issues are connected to some materials. A number of specialist volumes describes the health-related issues in some detail, including *Buildings and Health: The Rosehaugh Guide* (Curwell *et al.* 1990).

4.4.5.2   Are the skills and infrastructure needed to fit, install, and maintain available locally?

As well as having the potential to reduce or increase environmental impacts, the selection of construction materials is a basic catalyst for either increasing or reducing social and economic impact related to construction. A highly insulated, mass-produced, prefabricated building may well fulfil many of the virtues looked for within a sustainable building, but does this system of building, heavily influenced by the form of the materials it contains, really relate to the workforce in the area in which it is placed? Will the skills be available to construct the building, or will the labour have to be imported along with the factory-built materials? Will long-term maintenance contracts have to be agreed with providers a long distance from the building? Will the skills that are needed and the economic life blood associated with the use of those skills gradually develop over time, a natural process that will take time to complete? Should the design and implementation of the construction of buildings bear this in mind when choosing materials? In the same way, a new natural material that is locally produced may provide challenges for a workforce used to working with more processed materials and products that may not need careful or detailed design and handling. This is not to regard the use of either system, processed or natural materials, as preferable—both have many advantages when used in sustainable buildings—it is more a statement that new materials and products sometimes need to be considered in a broader context rather than simply assuming that any impact will be purely environmental.

4.5   Materials and recycle/reuse

The way materials are fitted, constructed, and maintained within a building are key to the possible options of recycling or reuse of components, or even whole building elements. Whilst an obvious preference is to reduce the amount of materials used in a building, it would be helpful at this point to identify the flow of materials from extraction to disposal and to clarify the terms *recycle* and *reuse*. Figure 4.2 shows the linear process that is normally associated with construction materials and products.

If materials within products are reused or recycled (including remade, renewed, refurbished, or remanufactured), then the linear process will have a built-in series of loops that will allow the materials to be diverted from waste disposal.

As shown in Figure 4.3 the linear trajectory of building materials to waste disposal is not the only option. Through the use of renewing, recycling,

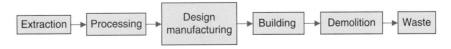

Figure 4.2 The linear process of material extraction to product disposal.

re-manufacturing, and reuse of building materials, much of a building can be diverted from landfill and make a useful contribution to future buildings. However, having a chart with a series of arrows indicating that materials can flow in loops away from waste disposal is much easier than actually getting this to work in reality. Whilst renewing and recycling of materials are possible on most construction sites, more energy can normally be saved and more materials directly utilised if the design team and supply chain can reuse materials and components.

Reuse of building materials is a challenge for any design team, whether the reuse of the materials is from the refurbishment of an existing building or pre-used materials taken from other places. This can be a worthwhile challenge as the finished building or structure can benefit from reduced emissions and raw material use. There is also possible added kudos and aesthetic appeal from a material that cannot be purchased off the shelf. However, the selection and location of reuse material, or complete products, is very different from the purchase of new materials or products. Reusing building materials, especially on a large scale for a new building, requires clear and direct communication between the design team and the contractors. This stems from five aspects that are not only important for new materials but also vital in relation to reused materials and components: dimensional fit and stability, acceptable aesthetic qualities, structural integrity, durability, and meeting safety standards. Through the use of the normal parameters that would be applied to any material and an emphasis on these five aspects, the contractor can communicate with the design team to narrow down the range of acceptable variance and communicate any deviations from those expectations.

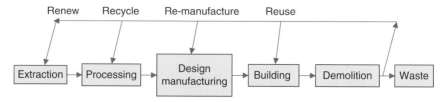

Figure 4.3 The closed loop process of the use of materials in buildings.

4.3 Want to know more? Reusing building components

If the reader wishes to obtain more specialist information concerning reusing building components, there are several possible references. These include BS 6543:1985, Guide to the Use of Industrial By-products and Waste Materials in Building and Civil Engineering (confirmed as current in 2013, BSI 1985). A more descriptive volume that has stemmed from a DTI Partners in Innovation Award, and that sums up many of the issues connected with building with reused materials, is Building with Reclaimed Components and Materials by William Addis of Buro Happold (Addis 2006).

or traditional brickwork. The specification of lime can often be justified where less impervious and more porous natural materials are used.

4.6.2 Alternative insulation materials

One way of reducing the amount of potential embodied carbon and, in some instances, the amount of toxins that might be associated with insulation materials manufactured from petrochemicals is by using natural insulation materials. Natural insulation materials made from fibres derived from hemp, wood, wool, cotton, cellulose, and flax have advantages and drawbacks when compared to other insulation materials. Natural insulations tend to be denser than their counterparts, leading to slightly higher thermal conductivities; therefore, to achieve the same thermal transmission (U) values, a great thickness of materials will be needed. However, this higher density can also be effective in reducing sound transmission and allow these materials to be used for two purposes. Natural fibres also tend to be less irritating to many people and therefore reduce the discomfort associated with placing rolls of fibre-based insulation in confined spaces. Any fibres that derive from plants or timber products may be a store of carbon, allowing the building to lower the building's carbon footprint, much as in the use of traditional timber. Many natural fibre insulations are also vapour permeable and are ideal for use in connection with construction details that encompass breathing solutions. This breathable nature does make this group of insulation materials unsuitable for use in positions that are vulnerable to damp without the protection of a damp-proof membrane. A helpful BRE publication, Information Paper 18/11 (Sutton *et al.* 2011c), introduces natural fibre insulation products and describes their properties alongside some comparisons with other insulations.

4.6.3 Hemp lime

A recent development based on the combination of a mix of chopped hemp stems and a particular mix of lime has resulted in a new construction material, hemp lime. Hemp lime has been used since the 1990s in France (Bevan & Woolley 2008) and follows a construction process of mixing a suitable lime (normally hydraulic lime or a formulated binder with some pozzolanic addition to increase set times) with hemp shiv (a chopped part of the woody internal core of the dried hemp stem after the fibrous outer part has been removed). The resulting non-load-bearing mixture is either sprayed onto a back board or more commonly poured between shuttering as infill between the structural members of a frame, normally timber. As the material does not offer a weather-resistant surface, hemp lime walling normally requires a weather resistance covering such as a rain-screen or render (preferably lime based) in a material that compliments the breathability that hemp lime offers.

As a walling material, hemp lime has a number of attributes that recommend it as a sustainable material. Hemp is a fast-growing material and in

Table 4.14 Summation of the findings of the BRE report concerning a comparison of performance of two hemp lime homes with masonry dwellings of the same size on the same site

Attribute	Hemp Lime Performance in Relation to the Masonry Homes at Haverhill
Embodied energy	Hemp lime accounts for 1.15kw/m³ and allows the hemp lime to have a lower embodied than conventional masonry materials.
Structure and durability	The qualities of hemp homes were found to be at least equal to those of traditional construction.
Thermal comparisons	Heating fuel consumed by the hemp homes is no greater than that used in the traditionally constructed houses.
Acoustics test	Hemp homes did not perform as well as the traditional houses but they did meet the sound resistance requirement.
Permeability	Both forms of construction appear to give complete protection against water penetration. However, the hemp homes generate less condensation.

comparison with timber provides a quick turnaround from field to form. Lime is produced at much lower temperatures than cement, reducing the overall energy needed for production. Lime takes on CO_2 as part of the carbonation process (see the section on Lime in this chapter) and can be seen as not only a store of the carbon from the hemp but also stores more carbon over time. The thermal properties of the material are conducive to keeping heat in during cold spells but also having enough mass to allow a structure to remain cool in high summer. The use of shuttering and filling a space between and around structural members reduces the probable waste by removing the need to cut that would be needed by masonry or other panel systems of infill. The ability of the hemp lime to wrap around structural members also minimises cold bridging.

The first researched hemp lime dwellings in the UK were a pair of two-storey properties in Haverhill in Suffolk by Ralph Carpenter of Modece Architects, compared with a series of adjacent masonry built dwellings by the UK's BRE. The external walls are 200 mm wide solid hemp lime, with a density of 550 kg/m³ with a 10 mm hydraulic lime render. The findings of the final BRE report (BRE 2002) are show in Table 4.14.

Hemp lime construction has also been used for commercial buildings, including the Adnams Brewery in East Anglia, the walling system used for the extension of the Lime Technologies industrial unit in Abingdon, Oxfordshire, and the WISE building at the Centre for Alternative Technology in Wales.

4.6.4 Straw bale building

Straw has often been considered a waste material in many countries, and to attempt to use such a material for construction purposes, if done appropriately, is sensible. Alongside straw's good thermal properties and ability to lock up carbon through a yearly supply, it has the potential to provide a sustainable walling material. The first commonly attributed straw bale buildings were built in Nebraska, where sources of timber and good quality stone was limited. These homes were often single storey with the straw bales acting as a load-bearing wall. Many modern straw bale buildings use a load-bearing frame, the straw bales acting as infill between and around the structural members. The straw bales are placed as a layer around the building on a suitable footing, with a method of fixing such as timber spikes to stop slippage on damp-proof membranes. The bales, then laid flat up to a height of 2.5 m, are compressed using straps and, depending on whether the walls are load bearing, the wall plate is either fixed (structural frame) or allowed to be lowered (load-bearing straw), with any gaps filled with appropriately strapped cut bales. Straw bale walls need to have a render finish applied to both the interior and exterior faces. Externally this can be in conjunction with a rain screen to provide added protection when the walls are subjected to higher than normal exposure ratings. Lime renders work well with straw bale buildings, allowing moisture to permeate, as do earth renders for interior coatings.

Whilst straw bale walls have distinct advantages, they also have a natural origin and require vigilance in respect to the variations associated with many natural materials. Bales need to be of sufficient density to be included in a wall, somewhere between 110 and 130 kg/m^3 and any restrung bales also need to be approximately in that range. Lower density bales could have lower thermal conductivities and improve thermal values further, but requirements of stability and fixing quality need to be taken into account. Straw bales on this density range should be perfectly acceptable when subjected to fire (Sutton *et al.* 2011a), if the prerequisite internal and external render layers are around 35 mm thick and complete.

4.6.5   Earth building techniques

Building using earth in its unfired form is not new; the use of straight-edged blocks dates back to 6000 BC (Lloyd & Safar 1945) and rounded earth blocks in the Middle East and Central Asia to possibly 1000–3000 years before that. In the UK some existing Cob buildings (a local form of earth construction) date back to around the 14th century. With this longevity and the relatively unprocessed nature of earth as a walling material, these types of buildings, if built and maintained appropriately, utilise one of the most sustainable construction techniques. The range of earth building techniques that could be chosen by a designer are broad, but all use subsoil of appropriate grading, enough moisture to allow compaction and cohesion, and in some cases chopped straw and a very limited amount of stabilizer such as lime cement or bitumen. Table 4.15 describes the characteristics of the various earth building techniques.

Some forms of earth construction have gone through a modern renaissance and are being used for a range of domestic and commercial buildings, either as the major walling material throughout or internally for performance reasons such as providing a form of thermal mass without exposure to the elements. The rammed earth walls in the main shop in the Centre for Alternative Technology, Mid Wales, UK, form part of the structure and allow visitors to view this construction technique close up (Figure 4.5).

Table 4.15 The characteristics of various earth building techniques

Earth Building Classification	Earth Building Technique	Characteristics
Monolithic earth construction	Rammed earth	Rammed earth walls are constructed in situ using a formwork or shuttering system, compacting the earth using mechanical or manual means with a relatively dry mix.
	Cob	Cob walls are not normally constructed using any shuttering and use a wetter subsoil/chopped straw mixture that is laid in layers or lifts approximately 300–400 mm high.
Earth brick or block construction	Adobe	Adobe walls are made from air-dried earth blocks, normally in a wooden frame and only hand compacted. A mortar of similar composition to the materials used to produce the adobe blocks is used in a similar fashion to building a traditional brick or block wall.
	Compressed earth blocks	These blocks have a similar composition to adobe blocks but are manufactured in a block-making press that can apply considerable pressure to the earth mixture, producing a denser block with different performance characteristics.

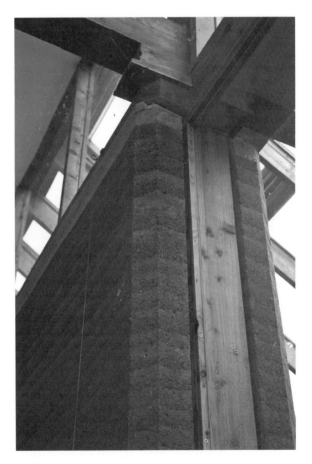

Figure 4.5 Rammed earth walls at the Centre for Alternative Technology, Mid Wales, UK (Photo by Steve Goodhew).

BSI (British Standards Institute) (2011) *BS EN 15978:2011 Sustainability of construction works. Assessment of environmental performance of buildings. Calculation method*, British Standards Institution.

BSI (British Standards Institute) (2013) *BES 6001 Responsible Sourcing of Construction Products: Helping You Prove Your Sustainable Building Approach* BSI, London. [Online], Available: http://www.bsigroup.co.uk/en-GB/bes-6001-responsible-sourcing-of-construction-products [6 Jan 2014].

Carbon Trust (2009) *Building the Future Today* Carbon Trust, London.

Curwell, S., March, C., & Venables, R. (1990) *Buildings and Health: The Rosehaugh Guide to the Design, Construction, Use and Management of Buildings* RIBA Publications, London.

EPA (2010) *Building Deconstruction and Design for Reuse, OSWER Innovation Pilot Results Fact Sheet* United States Environmental Protection Agency, Washington, D.C. [Online], Available: http://www.epa.gov/oswer/docs/iwg/building_decon_reuse.pdf [7 May 2014].

Green Construction Board (GCB) (2012) *Fundamental Truths* GCB, London. [Online], Available: http://greenconstructionboard.org/index.php/resources/fundamental-truths [01 May 2014].

Greenhouse Gas Protocol (2012) *Calculation Tools* Greenhouse Gas Protocol, Washington, D.C. [Online], Available: http://www.ghgprotocol.org/calculation-tools [6 Jan 2014].

Hammond, G., & Jones, C. (2011) *A BSRIA guide: Embodied Carbon—The Inventory of Carbon and Energy (ICE) BG 10/2011* University of Bath with BSRIA, iCAT, Berkshire. [Online], Available: https://www.bsria.co.uk/information-membership/bookshop/publication/embodied-carbon-the-inventory-of-carbon-and-energy-ice [6 Jan 2014].

HM Government (2010) *Low Carbon Construction Innovation & Growth Team: Final Report*. HM Government, London. [Online], Available: http://www.bis.gov.uk/assets/biscore/business-sectors/docs/l/10-1266-low-carbon-construction-igt-final-report.pdf [6 Jan 2014].

Holmes, S., & Wingate, M. (2000) *Building With Lime* ITP, London.

IPCC (2007) *IPCC Fourth Assessment Report: Climate Change 2007* IPCC, Geneva. [Online], Available: www.ipcc.ch/publications_and_data/ar4/wg1/en/tssts-2-5.html [6 Jan 2014].

ISO (2006) BS EN ISO 14040:2006 (Incorporating Corrigendum No. 1) *Environmental Management: Life Cycle Assessment — Principles and Framework*, BSI.

Lloyd, S., & Safar, F. (1945) *Tell Hassuna excavations by the Iraq Government Directorate of Antiquities in 1943 and 1944,* with prefatory remarks by Robert J. Braidwood. J.N.E.S. **4**, 255.

Recipro (2014) Recipro, Wirral. [Online], Available: http://recipro-uk.com [9 Jan 2010].

Royal Institute of Chartered Surveyors (RICS) (2012) *Methodology to calculate embodied carbon of materials RICS information paper (1st ed.)* RICS, London. [Online], Available: http://www.rics.org/Documents/Methodology_embodied_carbon_final.pdf [1 May 2014].

Sutton, A., Black, D., & Walker, P. (2011a) *Information paper: Hemp lime (IP 14/11)* IHS BRE Press, Watford. [Online], Available: http://www.bre.co.uk/filelibrary/pdf/projects/low_impact_materials/IP14_11.pdf [19 Dec 2013].

Sutton, A., Black, D., & Walker, P. (2011c) *Information paper: Natural fibre insulation (IP 18/11)* IHS BRE Press, Watford. [Online], Available: http://www.bre.co.uk/filelibrary/pdf/projects/low_impact_materials/IP18_11.pdf [19 Dec 2013].

University of Bath (2014) *Sustainable Energy Research Team* University of Bath, Bath. [Online], Available: www.bath.ac.uk/mech-eng/sert/embodied [6 Jan 2014].

USGBC (2007) Assessment of Technical Basis for a PVC-Related Materials Credit in LEED USGBC TSAVC PVC Task Group.

Woolley, T. (2013) *Low Impact Building: Housing Using Renewable Materials* John Wiley & Sons, Chichester.

Woolley, T., Kimmins, S., Harrison, R., & Harrison, P. (2000) *Green Building Handbook: Volumes 1 and 2 A Guide to Building Products and Their Impact on the Environment* Routledge, London,Oxford.WRAP (2014a) *Designing for Deconstruction and Sustainability* WRAP, Banbury. [Online], Available: http://www.wrap.org.uk/content/design-deconstruction-and-flexibility [9 Jan 2014].

WRAP (2014b) *Designing out Waste* WRAP, Banbury. [Online], Available: http://www.wrap.org.uk/content/designing-out-waste-1 [9 Jan 2014].

5. Construction-related sustainability

5.1 Sustainable construction

It is an unpalatable fact that building almost anything of substance that is meant to be permanent/semi-permanent, constructed and used on any piece of land will cause some form of disruption to someone or something. However, this first sentence is not intended to be a form of surrender to the inevitability of an unsustainable construction process, quite the opposite. Once the damage/potential damage, whether social, economic, or environmental, is pinpointed and discussed, valid practical solutions are not far away. Exemplars of good practice through major and minor companies working in the fields of construction can help other construction-related firms to reflect upon their work and do what most companies in the field are best at, solve problems and provide new solutions.

Once the design process has progressed to the stage when the construction activities can begin, certain issues that affect the sustainability of the finished product have been already fixed. The location, the building/structure's use and form, many of the specified materials, and who is going to undertake the construction will have already been decided. If it can be assumed that all efforts that could have been made have been undertaken to ensure the final product is as sustainable as can be expected, these can still be undone in either major or minor ways by 'getting it wrong' on site and after the building is occupied. This chapter examines the issues that link sustainable construction and the process of building, managing the facility that has been built, and dealing with the process of demolition sustainably.

One of the greatest issues concerning the sustainability of the construction process is the variability of the many means and procedures that it takes to build a building or structure. The context of the construction can also influence the ability to judge whether one process is more or less sustainable than another. Jia Sheng Jin quite rightly comments that 'Green construction is a relative concept' and that the content of any technical innovations will vary from country to country, region to region when related to local interpretations

Sustainable Construction Processes: A Resource Text, First Edition. Steve Goodhew.
© 2016 John Wiley & Sons, Ltd. Published 2016 by John Wiley & Sons, Ltd.

of the social, economic, and environmental aspects of sustainability (Sheng 2013). Whilst the local conditions may influence the interpretation of the general precepts of sustainable site operations, there are many common threads that can ensure that the construction process is sustainable.

To ensure, fairly, that the building or structure has been produced using the most sustainable practices is a complex task; this chapter examines the aspects of construction process and sustainability to match the two and offer guidance.

Appropriately, it is first necessary to categorise each of the broad aspects of the construction process and their relationship with each of the aspects of sustainability (Table 5.1).

Table 5.1 The broad aspects of the construction process and their relationship with each of the aspects of sustainability

Sustainability-Related Aspect	Construction-Related Process/Impact
Energy in the process	Virtually the whole process, from site clearance and ground works right through to maintenance of the finished product
Materials	A combination of the materials, both temporary and permanent (but not chosen through the design process), that are used in the construction process. The materials that are used during the maintenance of the building or structure and issues concerning how any unwanted materials are dealt with when the building is demolished or refurbished.
Water	Many processes are reliant on good-quality water. Some processes actually consume the water, such as wet finishes, whilst others such as washing tools, etc. could offer some potential reuse.
Health and safety	Building and other structures that employ people in temporary and potentially dangerous situations offer a duty of care to employees and bystanders. This directly relates to the social element of sustainability. Particular care is taken when demolishing buildings and structures.
Physical impacts on the region immediately surrounding the project	Noise, dust, traffic flow, the aesthetic and the security of the site
Economic impacts on the region immediately surrounding the project	Local employment, use of local suppliers (both where appropriate), linking with local authorities when the project type and size is appropriate
Economic, social, and environmental impacts on the region once the project has been completed	The employment of staff to maintain, clean, and run the building or structure. The choice of systems to ensure that the building services and the materials used through the life of the building are benign as far as the health of the occupants and the surroundings are concerned.
Beyond use issues	Demolition, recycling, and reuse of parts or the whole building

5.2   Site operations and organisation

Construction site operations and their relationship with sustainable construction have for some time been seen to be connected, but constraints related to the more traditionally perceived priorities of what construction site management are have hampered the introduction of more sustainable practices. According to Robichaud and Anantatmula (2011), 'In order for project managers to deliver sustainable construction according to clients' cost expectations, modifications must be made to traditional project management processes and practices'. Different aspects of improving the sustainability of the construction process at a site level have been studied by a number of researchers and organisations. Xin Zou and Sungwoo Moon (2013) have evaluated the efforts to improve environmental performance of site-based processes and concede that these efforts 'should be aimed not only at operating and maintaining built facilities but also at constructing them'.

Mollaoglu-Korkmaz and colleagues (2013) investigated the links between project delivery methods and sustainability goals through twelve case studies, finding, 'The results show that the level of integration in the delivery process affects final project outcomes, particularly sustainability goals'.

A similar case study approach was undertaken by Halicioglu and others (2012), alongside an analysis of the experiences of project managers involved in building projects. The findings concluded that many issues, including technical, environmental, economic, social, organizational, and innovation-related ones, can be central to the success (or not) of sustainable building projects.

5.2.1   Building Information Modelling

Site operations are likely to be influenced by the introduction of Building Information Modelling (BIM). BIM is discussed in more detail in Chapter 6, but suffice to state that it is a method of sharing knowledge resources to support

---

5.1 Want to know more? The management of sustainable construction

Exploring the management of sustainable construction at the programme level: A Chinese case study, *Construction Management and Economics*, Volume *30*, Issue 6, 2012, pages 425–440, Qian Shia, Jian Zuob, & George Zillanteb. DOI:10.1080/01446193.2012.683200

Criteria for the selection of sustainable onsite construction equipment, *International Journal of Sustainable Built Environment*, M. Waris, M. Shahir Liew, M. Faris Khamidi, & Arazi Idrus, Volume *3*, Issue 1, June 2014, pages 96–110, ISSN 2212-6090, [Online], Available: 10.1016/j.ijsbe.2014.06.002.

http://www.sciencedirect.com/science/article/pii/S221260901400034X [31 Aug 2015].

decision making about a building project. This support is intended for the very early conceptual stages, through the detailed design phase, the construction operations, and beyond to aiding people through the building's operational life and finally demolition/end of life. Whilst there is limited current evidence of BIM being used in documented case studies (Davies *et al.* 2013), Mäki and Kerosuo (2013) followed two case studies of renovation sites in Finland, focusing upon the day-to-day use of BIM by site managers. As BIM becomes more widely used in the UK past the government's requirement for its compulsory use in 2016, the linkages to sustainable construction are likely to become more apparent. The feedback from design, orientation, and specification changes will be more accurate and be communicated more quickly. This increase in effectiveness and likelihood of higher quality shows that BIM is increasing the ability to achieve designs that would be impossible without digital design and fabrication. This can be extrapolated as a contribution to sustainable construction (Rajendran *et al.* 2012).

5.2.2 Energy

Minimising the use of energy or the influence of the site processes upon the future energy use of the building and structure are part of ensuring that the construction process is indeed sustainable. This description of energy

---

5.2 Want to know more? Building information modelling (BIM)

BIM further reading:

Dossick, C.S., & Neff, G. (2010) Organizational divisions in BIM-enabled commercial Construction, *Journal of Construction Engineering and Management ASCE*, 459–467.

Eastman, C., Teicholz, P., Sacks, R., & Liston, K. (2011) *BIM Handbook: A Guide to Building Information Modeling for Owners, Managers, Designers, Engineers, and Contractors*, John Wiley and Sons, Hoboken, New Jersey, US.

Hardin, B. (2009) *BIM and Construction Management: Proven Tools, Methods, and Workflows*, Wiley Publishing, Indianapolis, Indiana, US.

Likhitruangsilp, V., Putthividhya, W., & Ioannou, P. (2012) Conceptual Framework of the Green Building Information Management System, *Construction Research Congress 2012*: pp. 658–667.

UK BIM resources:

BIM UK Strategy (Innovate UK 2015): https://connect.innovateuk.org/web/process-efficiency/bim

Digital Design and Engineering (DGE 2015): https://connect.innovateuk.org/web/digital-design-and-engineering/overview

Digitising the Construction Sector: Launch Presentations (DCS 2015): https://connect.innovateuk.org/web/digital-design-and-engineering/bim

could be as simple as the fuel used in site vehicles or more hidden, such as the choice of a particular product that has more energy embodied within it than an alternative product.

5.2.2.1 Site-based energy use

From Table 5.2 it can be seen that the energy use and greenhouse gas emissions that can be attributed to construction processes in North America are significant.

These significant amounts of energy use also come from UK site-based construction operations (Table 5.3). The UK construction statistical indicators (KPIs, key performance indicators) are amassed by Glenigan (2012) on behalf of the UK Government and can be accessed via the UK Government's Office for National Statistics, http://www.ons.gov.uk/ons. The KPIs have a section relating to the environmental performance of the construction industry, including energy use, a UK energy performance indicator (the Standard Assessment Procedure for the Energy Rating of Dwellings), designed average waste, creation or retention of habitat, onsite energy use, mains water usage, waste removed from sites, and commercial vehicle movements.

As can be seen in Figure 5.1, whilst the site energy usage has dropped in the UK for all construction work, there is a very clear correlation between the amount of construction output and the variations in energy used by construction sites. One can therefore continue to assume that reductions in site-based energy use are still possible as the levels of construction output increase in the coming years. WRAP (2013a) estimates that 'straightforward and easy to implement energy efficiencies' can provide typical cost savings of 15%. To understand some of the site-based processes that should be targeted for energy reduction, it is helpful to use an analogy. Producing a building using a large team of people in a safe and sustainable way is similar to the flows of energy, materials, and labour of a factory, complete with buildings and canteen, being moved to a temporary location to build a 'one-off' product. It is therefore fair to assume that looking after the health and welfare of everyone involved in the construction process could end up requiring as much energy as it does in an equivalent factory. This site-based energy is added to the totals that are described as the embodied energy of the complete structure and can be approximated as CO_2/£100k project value, shown in Table 5.4.

Table 5.2 Energy use and GHG (greenhouse gas) emissions from construction sectors in the United States and Canada in 2006.

Country	GDP (nominal billion $)	Share in National GDP	Energy Consumption (trillion Btu)	Share in National Energy Consumption	GHG Emissions (Tg)	Share in National GHG Inventories
US	649.4	4.90%	913.9	1.2%	67.2	1.2%
Canada	75.4	5.95%	57.5	0.7%	4.2	0.9%

Sources: USBEA 2014; StatCan 2014; EPA 2015; NRC 2014; Fergusson 2008; NRCAN 2008.

Table 5.3 A comparison of the construction site energy use since 2003 and the construction output for the UK, all work

Construction Process Performance	2003	2004	2005	2006	2007	2008	2009	2010	2011	2012
Energy use (current values $CO_2/\pounds100k$ project value)	288	322	293	293	273	192	241	249	267	196
Energy use (2005 values $CO_2/\pounds100k$ project value)	242	288	273	293	287	213	278	286	300	223
Mains water use current values ($m^3/\pounds100k$ project value)	7.5	9.7	8.2	8.9	8.2	7.1	6.3	6.3	4.9	6.9
Mains water using 2005 values ($m^3/\pounds100k$ project value)	6.3	8.7	7.7	8.9	8.6	7.9	7.3	7.2	5.5	7.9
Waste current values ($m^3/\pounds100k$ project value)	43.5	47.1	41.6	37.0	39.1	36.9	36.6	35.1	26.7	19.4
Waste 2005 values ($m^3/\pounds100k$ project value)	36.6	42.2	38.8	37.0	41.2	40.8	42.2	40.4	30.0	22.1
Commercial vehicle movements, current values median movements onto site per £100 project value	44.0	34.5	29.4	30.4	29.4	26.5	28.3	23.1	19.7	49.5
Commercial vehicle movements, 2005 values median movements onto site per £100 project value	37.0	30.9	27.4	30.4	31.0	29.3	32.6	26.6	22.2	56.4
Construction output all sectors, to 2005 values, all work 4th quarter figures	101.3	101.1	98.8	102.6	102.9	93.7	86.3	94.9	95.6	87.1

Sources: Office for National Statistics 2013; Glenigan 2012.

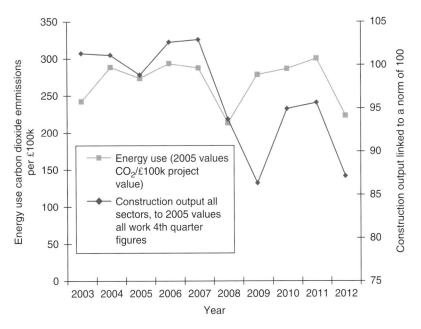

Figure 5.1 Energy use on construction sites compared to the construction output between 2003 and 2012.

Table 5.4 Types of operations that use energy through site-based activities

Type of Energy Use	Details of Energy Use
Transportation of materials to site	The energy that is counted in this category has to be related to site processes, so the further transport energy involved in, perhaps, imported goods from overseas would qualify for the more general category of embodied energy. However, where there is a choice of materials such as ready-mixed concrete, this can legitimately be regarded as contributing to site-based energy.
Powered site construction operations stemming from mobile plant and equipment	An example of these are tools activated by compressed air that are fed by mobile compressors normally run by a diesel internal combustion engine. Electric tools such 'breakers' and 'chippers', drills, saws, and staple guns are also add to the energy tally.
Fixed plant	Energy used by fixed plant not readily transportable off-site in a short period of time.
Site operatives and management-related energy	The travel to the site, PCs needed to run the site, photocopying, faxes, large-scale drawing printing
Site accommodation and related services	Heating for site accommodation, bathing and toilet–related energy use; site canteen (if applicable)

Site-based energy usage comes from a variety of operations and reasons, some of which are involved with all construction processes and others which are unique to particular projects. Table 5.4 shows some of the general groupings of site-based energy use.

BREEAM (the Building Research Establishment Environmental Assessment Method, further details in Chapter 6) Section 4 Man 03 gives credits in relation to measures that are designed to reduce site-based energy consumption. One credit is given when the energy used in kWh by the site processes that produce the building or project is monitored and data recorded for energy consumption (kWh) from the use of construction plant, equipment (mobile and fixed), and site accommodation (assuming that all these are necessary for the execution of all construction processes). The guidance in Man 03 also requires the following: 'Using the collated data, report separately for materials and waste, the total fuel consumption (litres) and total carbon dioxide emissions (kgCO$_2$eq), plus total distance travelled (km) via the BREEAM Assessment Scoring and Reporting tool' (BRE 2015). This is an obvious starting point to allow the valid and appropriate management of energy, water, and waste-related site-based resources. However, to allow the monitoring to pinpoints areas of excessive energy consumption, three further elements are required:

1. The use of submetering to help identify the type of consumption that needs to be reduced
2. Some past data or estimated/suggested quantities of energy that might be used when best practice is undertaken and then used as comparator KPIs
3. The ability to record the energy data with a fineness of resolution that will allow any waste to be identified and managed. This might include energy usage when the site is closed, such as over vacation periods or use after working hours.

One development that will assist a site manager to record energy and other BREEAM Man 03 data is a smart-phone app that BREEAM has produced that assesses both electricity and water meters per site, can set KPI targets per meter, and will allow the import of existing readings from a CSV file alongside the ability to view graphs of monthly usage.

It is likely that as the need for site-based energy monitoring is driven by guidance and expectations, the electronic and online products, both software- and hardware-based, will develop rapidly.

5.2.2.2 Lighting in site accommodation

Except for the most up-to-date site accommodation, many site offices, canteens, and welfare cabins are basic and dated. Their lighting facilities are no exception. The use of traditional fluorescent tubes ensures that the overall energy use on site is higher than most of the more modern systems. According to a study carried out by a masters student at Strathclyde University that focused on the modelling of energy use on a construction site (Ndayiragije 2006), replacing the standard fluorescent tubes (T8 58W) with the more energy-efficient fluorescent tubes (T5 49W) would reduce CO$_2$ emissions by 27 tonnes per year and save £5401 per year. This does seem a little optimistic, and some of the assumed savings due to correctly functioning control systems might not be forthcoming; however, even a reduced reduction would still be welcome.

Table 5.6 Various possible sources of construction-related noise and the stages of development of a building or structure

Stage of Construction/ Building Use	Potential Source of Noise
Site clearance	Earth-moving machinery, chainsaws for vegetation clearance
Ground works	Machinery needed for creation of foundations, waste soil collection vehicles
Main construction	A number of internal combustion engine vehicles, electrical and pneumatic hand tools, skip collections
External works	Road and path making operations, digging, rollers, and compaction devices
Maintenance during occupation/use	Cherry pickers and other hoists for painting, recladding, etc.
Reuse/demolition	A large range of vehicles augmented by wrecking balls, hand demolition, and even explosives

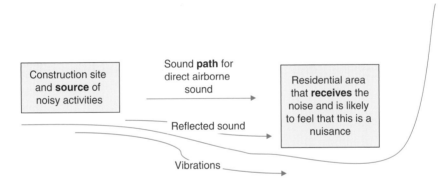

Figure 5.2 Diagram of sound source/path and receiver.

equipment and limiting the amount of time that an operative might use such equipment. The latter issue has a direct nuisance aspect and touches directly on the lives of those who live and work in reasonable proximity to a site operating heavy plant. The level of vibration will vary with the type of plant used and the soil conditions on and under the site and hence the level of nuisance caused. External works or foundation-related operations are two of the most common site-based operations that are likely to come under this category. As with some of the methods of mitigating disturbance due to noise, one option is to change the method of operation, prevention rather than cure. Using bored pile rather than driven piles is likely to reduce the levels of vibration caused as well as the noise generated. One issue relating to the transmission of vibration is the distance to the nearest building, and a study found that nuisance from ground vibration and building damage is unlikely to occur if the operation is conducted at distances greater than 50 metres (Martin 1980).

Table 5.7 Range of potential full or partial solutions to some commonly occurring noisy situations

Source of Noise	Potential Solution at Source	Potential Solution for the Path of the Noise	Potential Solution at the Position of the Receiver
Site operations connected with noisy hand tools	Rethink the method of undertaking the operation: can a quieter method that perhaps uses a non-impact solution be used?	Screening either around the work area or as a less effective solution screening further from the operation	As in many areas of sustainability-related solutions, dealing with an issue at source tends to be far more effective than dealing with the issue at a later stage. Noise control is no exception and prevention of the noise at source is normally better for all parties. However, here are a few potential 'cures' for the receiver.
Internal combustion engine–related noise	Possible replacement with a quieter method such as an electric or hydraulic power source. Renew/check silencer or other noisy parts.	Encasement of the engine with an insulating box, with particular attention to sealing at joint positions	
Ground work–related noise such as piling	Consider using an alternative method of foundation design if feasible. If piles are required, then consider bored piling solutions rather than noisier driven piles.	Screening can sometimes be partially effective, but the percussive nature of driven piles often means that restricting the periods of noise-related activity is the more appropriate solution.	Extra insulation or modification of the potential sources of sound 'leaks' in the facades of surrounding buildings, such as wide-spaced secondary glazing, and acoustic baffles on vents. Screening close to the receiver, preferably of aesthetically appropriate construction and of appropriate materials.

5.2.5   Dust

In future construction operations there may come the day when such a wide range of the traditional site-based activities are based off-site in factory conditions. The dust associated with many of the current activities on a typical construction site could be gathering under controlled conditions. The impacts of dust on site staff, local inhabitants, and workers do have health, nuisance, and pollution implications. Many related dust site-based activities, such as ground works; external works; acts of cutting, grinding, and site cleaning; and clear ups will still have to take place even if modern methods of construction are far more ubiquitous than they are today.

5.2.5.1 *The constituents and sources of dust and other airborne related pollutants*

Fine particles can be classified as dust. The Control of Substances Hazardous to Health Regulations 2002 (COSHH) require workers in all work place situations to be protected from the harmful health effects of hazardous substances, including dust.

There are three main categories of construction dust:

Silica dust is present in large amounts in construction materials such as most concretes, mortars, granite, sandstone, and sand-based fine aggregates. According to the UK's Health and Safety Executive (HSE 2013), 'The silica is broken into very fine dust (also known as respirable crystalline silica or RCS) during many common tasks such as cutting, drilling, and grinding'.

Materials such as marble, gypsum, and limestone produce **non-silica** dust when worked. Dust from excavations is most likely to be non-silica in nature. Logically the machining of timber on construction sites produces **wood dust** from softwood, hardwood, and a range of manufactured wood–based products such as chipboard, plywood, and M/HDF (medium/high density fibreboard).

The reasons for endeavoring to restrict the amount of dust either being produced or being transported within or without the site boundaries is three-fold. The dust is a health hazard. Studies have linked working in a dusty environment to various respiratory diseases such as asthma or bronchitis (Becklake 1985; Meredith *et al.* 1991). Dust can cause situations where for safety reasons the carrying out of operations becomes more risky, including increased risk of tripping or slipping of both people and machinery. Dust can also be harmful to the operations being undertaken, covering up defects, becoming mixed with other substances and generally coating site-related items with dust, reducing efficiency, and sometimes making quite basic operations difficult.

The sources of dust are quite varied but do stem from a number of identifiable tasks, and so dust reduction can hone in on these general classifications of tasks. Any work that is connected with the soil that the building is placed on and the roadways, either temporary or permanent, will in the summer months give rise to the production of dust. The amount of dust and the fineness of the dust particles is likely to be related to the soil on the site. Operations involving cutting of materials will give rise to quantities of dust. Angle grinders, often due to the high speeds of rotation of the discs used, will tend to produce finer dust than traditional manual tools such as saws. The materials being cut should be adequately assessed for risks from the potential dust being produced. If any dust is likely to be harmful, then off-site cutting should be built into any method statement linked to the overall operation. Some sources of dust are not related to site-based manufacturing operations but to the powder materials that form an intrinsic part of present aspects of construction practice. These include plasters, cements, fine sands, grouts, and powder-related finishes. Many of these are being reduced by off-site processes or practices, such as using dry lining internal boarded finishes and ready mixed concrete rather than using site-based batching

Table **5.8** Specific reduction strategies for dust reduction on construction sites

Source of the dust	Specific reduction strategy	General reduction solutions
Cutting	Avoid cutting by using precut or modular systems or prepared off-site components.	Consider alternative construction methods or technologies that will eliminate cutting, powdered materials, and minimise soil-related operations.
	Use liquid to reduce dust production for compatible cutting/drilling operations.	
Earthworks/ soil-related dust	Limit the numbers of dig sites and where possible keep areas of digging to manageable positions, grouping them so that the dust can be confined to one or few sections of the site.	
	Using hosing and damping down to reduce the travel of dust to a minimal amount.	
	Collecting rainwater to use for dust reduction purposes.	
Powder and granular construction materials	Maintain good materials management procedures accounting for the position and quality of the storage for any materials that are liable to cause dust. Ensure that the movement of bagged powder and granular materials considers the fragility of the bags and the appropriate carrying of heavy loads.	

plants. This, to a great extent, has removed the powder-related dusts, but final finish skim coats for plasters and 'making good' defects in poured or precast concrete is still going to be a source of site dust through the admittedly smaller quantity of powdered materials needed for these tasks.

5.2.5.2   Dust solutions

As with many of the sustainability-related construction site issues, prevention tends to be a better strategy than cure, as the resulting solution is often more effective and efficient. Some elements of this prevention are shown in Table 5.8.

5.2.6   Water

As the attention spent on energy use and waste on construction sites starts to produce tangible reductions, other aspects of the construction process also come under scrutiny. Water use on construction sites is then quite rightly seen as an area of activity where little is known about the details of water use but logically requires focus. Part of this focus is provided by the changing rainfall patterns (Aqueduct 2014) and the UK's *Strategy for Sustainable Construction* (HMS Government & Strategic Forum 2008),

Whilst it could be argued that construction can have a potential impact on all of the above, the most frequently affected are breeding farmland birds, butterflies of the wider countryside on farmland, plant diversity in neutral grassland and boundary habitats, widespread bats, breeding woodland birds, and butterflies of the wider countryside in woodland. This document focuses on biodiversity rather than impacting activities, and when the search function is used within the document, the term *construction* gets zero 'hits'. It is therefore necessary to look for guidance from more construction-related sources.

Two documents that have bearing on biodiversity and construction come from two construction-focused groups. In 2003 The UK's BRE and CIRIA developed the following set of biodiversity indicators that allow the impact of construction projects on biodiversity to be measured: (1) Impact on biodiversity: product, (2) Impact on biodiversity: construction process, and (3) Area of habitat: biodiversity indicators for construction projects (Woodall & Crowhurst 2003). The 2008 UK 'Strategy for Sustainable Construction' (BERR 2008) has a section (10) that is devoted to the links between construction and biodiversity. This section's overall aim is that 'the conservation and enhancement of biodiversity within and around construction sites is considered throughout all stages of a development'. To achieve this the strategy set out a number of actions and deliverables focused upon this area that include the following:

- All construction projects over £1 m to have biodiversity surveys carried out and necessary actions instigated
- Biodiversity Toolkit for planners and local biodiversity officers
- The UK Green Building Council to hold workshops and set up a task group to develop a roadmap for the industry to maintain and enhance biodiversity

The Biodiversity Toolkit for UK planners and local biodiversity officers has been completed and is available at http://www.biodiversityplanningtoolkit. com/default.asp (ALGE 2011). This toolkit is an interactive series of tools and allows the viewer to look at sections relating to subjects such as

Fundamental considerations
Law/policy/practice
Types of development designations
Key species
Key habitats
Key geodiversity
Survey calendars
Forward planning
Interactive bat protocol

The UK's BRE's Environmental Assessment Methods (BREEAM) (see Chapter 6) cover a range of non-domestic building types, including offices, industrial buildings, schools, and multiresidential developments.

BREEAM assessments give credits for protecting existing ecological features and for ecological enhancement such as green roofs on new buildings. The assessment does require an ecologist to recommend enhancement measures that contribute to Biodiversity Action Plan targets and to promote best practice among contractors and asset managers. This is an assessment, much as LEED is in the USA, that can be used by any interested body or individual to assess and help improve the performance (in this case for biodiversity) of a building and its surroundings. Alongside these assessment systems individual companies and institutions are tailoring their own plans or schemes to enhance their own buildings and sites. Marks and Spencer, the large UK-based clothing and food store, has a scheme entitled 'Plan A: Doing the Right Thing', which introduces biodiversity audits and action plans for major store building projects from 2011. They state, 'Our aim is to create new stores with a net positive biodiversity impact which means that the level of onsite biodiversity post-construction is higher than it was before we developed the site' (Marks & Spencer 2010).

Ecological surveys can provide the necessary baseline information for the creation of a site biodiversity action plan, which through the guidance of suitable sources such as English/Scottish/Welsh Nature can focus upon matching the needs of the site and impending construction processes.

In the future more linkages between individual construction projects and reducing any impacts on ecosystems might be founded on a possible monetary value of good levels of biodiversity. Comello, Lepech, and Schweglar have researched reasons that prevent people involved in a construction project from considering environmental concerns (Comello et al. 2012). They put this down to the lack of (1) a clear understanding of the connections between project ecological effects and how these damage firm/project assets and (2) the ability to include the project ecological effects in operational decisions. These authors further present a framework for ecosystem service valuation that consists of four parts:

1. Life cycle assessment to quantify project ecosystem emissions;
2. Fundamental biophysics and biochemistry to characterise the component processes of ecosystem services;
3. Functional substitutability to assign a monetary value to such services; and
4. Representation of ecosystem services value within international financial accounting norms.

Whilst winning credits in assessment exercises and allowing some corporate entities to use the policies and actions related to a construction project for marketing, the use of monetary equivalents may well prove to be one way of increasing the reach of sustainable construction in relation to biodiversity.

5.2.8   Pollution (general)

With the need for construction also come associated processes that will lead to the risk of pollution. In this instance pollution would be defined as a release of a substance that can cause harm to people, animals, plants, air, water, or soil.

The environmental agency of the UK has a considerable amount of advice collected together in PPG6. This guidance is aimed at site managers of large, medium, and small-to-medium enterprises and attempts to introduce methods of preventing pollution for construction sites.

According to PPG6 the most common pollutants that come from construction processes are silt, oil (including fuel), cement, concrete, grout, chemicals, sewage, waste materials, dust, and smoke (also including exhaust fumes).

5.2.8.1   Soil

The generation of soil takes many years (up to 500 years according to some sources) and relies on factors such as the parent material, the climate, topography, and organisms in the soil (Jenny, 1941). This along with other drivers implies that soils are not a renewable material when considered within reasonable timeframes under natural conditions and reinforces the importance of retaining the properties of the soil on construction sites. The functions of soil include

- supporting plant growth (food production),
- recycling nutrients and waste,
- controlling the flow and purity of water,
- providing habitat for soil organisms, and
- functioning as a building material/base.

(Covert 2013)

Some of the ways in which soil can be degraded due to construction process activities include contamination, overcompaction, reducing soil quality, losing soil by allowing it to be mixed with waste materials that are taken from the site and disposed of, and sealing of the surface to the soil, resulting in detrimental impacts upon the soil's properties.

Whilst in the UK there are no specific direct planning controls on the sustainable use and management of soil resources on construction sites, there are steps that can be taken and guidelines that can be followed to limit soil-related impacts. The *Construction Code of Practice for the Sustainable Use of Soils on Construction Sites* published by the UK's DEFRA (2011) has detailed sections relating to good practice. This document also summarises a sequence for contractors or site operations that is designed to lead to the sustainable handling of soil.

5.5 Want to know more? Soil and preconstruction planning

What should be scheduled and planned to maintain the quality of soil resource?

- Have a soil resource survey carried out on site by a suitably qualified and experienced soil scientist or practitioner (e.g., a member of the Institute of Professional Soil Scientists, www.soilscientist.org) at the earliest convenience and prior to any earthworks operaions.
- Incorporate the results of the soil resource survey into the site working strategy (e.g., site waste management plan or material management plan), ensuring liaison between the soil resource survey and other ground investigations.
- Ensure that you are informed of and follow waste regulations as necessary.
- Consider the use of sustainable drainage systems on site as these can provide more long-term protection of soils beyond the construction phase, by facilitating the infiltration and attenuation of surface water.

Soil management during construction

- Prepare a soil resource plan showing the areas and type of topsoil and subsoil to be stripped, haul routes, the methods to be used, and the location, type, and management of each soil stockpile.
- When stripping, stockpiling, or placing soil, do so in the driest condition possible and use tracked equipment where possible to reduce compaction.
- Confine traffic movement to designated routes.
- Keep soil storage periods as short as possible.
- Clearly define stockpiles of different soil materials.

Landscape, habitat, or garden creation

- Ensure that the entire soil profile is in a condition to promote sufficient aeration, drainage, and root growth.
- Safeguard and utilise on-site soil resources where possible. If importing soils, use a reputable supplier, establish the source of the soil, and ensure it is suitable for the intended use.

DEFRA (2011) *Construction Code of Practice for the Sustainable Use of Soils on Construction Sites*, https://www.gov.uk/government/publications/code-of-practice-for-the-sustainable-use-of-soils-on-construction-sites

Figure 5.3 Rammed earth wall as part of a complex of buildings in Lincolnshire UK (Photo by Steve Goodhew).

construction sites, but the material does not react well to some practices that would be judged acceptable when used in conjunction with traditional masonry construction. In the UK earth blocks can be a good solution to use for internal partitions where the levels of protection from external moisture are not required. Ibstock has manufactured an unfired earth brick in the UK that can be ordered commercially and delivered alongside appropriate mortars and finishes for use in both domestic and commercial buildings.

5.3.3 Operational waste, modularization

Operational waste has a direct relationship to waste that occurs due to design decisions. Section 4.5, 'Materials and Recycle/Reuse', in Chapter 4 discusses these aspects along with the wider practical and theoretical aspects of the ability to design for reuse/recycling. This section attempts to focus on the site and other operational processes.

Figure 5.4 Series of cob walls built as part of a 3-storey large cob dwelling in Devon UK (Photo by Steve Goodhew).

5.3.3.1   Demolition

As the focus on brown field sites continues, the importance of the good management of materials when demolishing existing structures is brought to the fore. Demolition, a process that was in the past (and in some ways continues to be) purely a destruction process is now directed towards a method for recovering reusable and recyclable materials alongside appropriate disposal of any remaining waste. This mindset in itself can promise more sustainable solutions and better decision making. The safety and responsibilities for demolition were laid down in the UK by BS 6187:2000, The Code of Practice for Demolition. BS 6187:2011 has now been revised to reflect changes in the construction and other industries and advises on the following three areas:

- The deconstruction of buildings and structures, including activities for reuse and recycling
- Managing a project, including site assessments, risk assessments, decommissioning procedures, environmental provisions, and façade retention
- Safe working, including elements of design and application

The standard also raises the issue of data recording and collection via electronic means. This enables contractors and other interested parties to potentially access information about materials and products that could be reused and recycled via exchange websites such as SMARTWaste and BREmap. In a world where BIM is offering many new opportunities for sustainable construction, embedding this form of information in a BIM materials map could afford better future planning in relation to demolition.

5.3.4 Construction waste

Construction waste is one of the major impacts of the construction process, both in the UK and across the globe. According to Yeheyis *et al.*, 2013, the construction and demolition waste generated by the Canadian construction industry accounts for 27% of the total municipal solid waste disposed in landfills. These authors also state that over 75% of what the construction industry generates as waste has a residual value, and therefore could be recycled, salvaged, and/or reused. As can be seen from Figure 5.5, the waste attributable by the UK construction sector forms 34% of the total waste generated.

Of the 34% shown in Figure 5.5, over 25 million tonnes of construction, demolition, and excavation waste ends up in landfill every year (BERR 2008).

In an attempt to reduce this large amount of waste, site managers can follow a series of proven techniques that have been applied to other spheres of management. This will often involve using monitored data and linking it to targets of a taxing but achievable size and scope. The settings of those targets can be a difficult task without reliable reference data. To establish these targets the current wastage needs to be measured and future performance benchmarked against an evolution of these. Actual measurements of site waste can vary considerably depending on the type of building or project being constructed. Standard figures used in pricing books can provide one method of establishing waste rates, but to allow construction managers to aim for best practice the UK's Building Research Establishment issued a series of average waste benchmark data (Table 5.10) through www.smartwaste.co.uk. Table 5.10 shows these

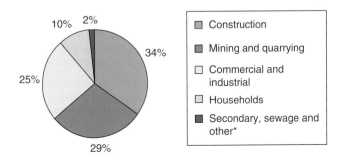

Figure 5.5 Total UK waste generation by sector, 2004 to 2008 (UK Government 2011).

Table 5.10 Average UK waste benchmark data, issued 26 June 2012

Project Type	Average m³ per 100 m²	Average Tonnes per 100 m²
Residential	18.1	16.8
Public buildings	20.9	22.4
Leisure	14.4	21.4
Industrial buildings	13.0	12.6
Healthcare	19.1	12.0
Education	20.7	23.3
Commercial other	17.4	7.0
Commercial offices	19.8	23.8
Commercial retail	20.9	27.5

figures for a range of project types leading up to 2012. As the overall construction waste tonnage is reducing over time, it would be logical that these figures should also fall over the coming years, fulfilling targets within the 2008 UK sustainable construction strategy and the expectations of clients and the industry in general.

A number of different investigations have been undertaken to research the relationship between construction site operations and most aspects of site waste. A recent piece of work undertaken by Yates (2013) investigated the main types of construction wastes and sustainable strategies that could be used to minimise the amount of waste generated by the construction industry. This work is important as it focuses on both the building construction industry and more engineering-related projects. Good practice found in one branch of construction might well be applicable to another. However, this work does tend to rely on expert statements from high-level executives relating to strategy, and this might allow a view of more site-based practices to be, to an extent, lost (Yates 2013). A recent paper by Hongping Yuan (2013) takes a holistic perspective of construction waste management and identifies 30 key indicators that can be used to improve the effectiveness of site operations.

5.3.4.1   Site waste management plans

Up until 1 December 2013 all construction projects over £300,000 in England needed to have a site waste management plan (SWMP). Currently (at time of writing), the UK government is encouraging contractors to still utilise SWMPs but under the auspice of flexible resource efficiency tools rather than via legislation. In the SWMP a contractor or responsible person (the person responsible for the plan is the owner of the site, but this is more usually delegated to the contractor, site manager, or project manager) must explain how construction waste is handled and how the legal aspects of managing waste and waste disposal will be adhered to. Previously in England, it was not possible to start any eligible construction work until an SWMP was in place.

The SWMP should include

1. what kind of waste the site produces;
2. how the waste is disposed of, for example, reuse, recycle, landfill;
3. who the waste carrier is and their registration number;
4. the address and environmental permit or exemption number of the site where the waste is being taken.

The plan would be the active responsibility of the main contractor, with a view that the contractor would then take ownership for the plan, adding any waste information from the subcontractors. It is then very important that the plan becomes 'live' and that the main contractor makes sure everyone working on the project knows how to dispose of waste as laid out in the plan. The plan also can become out of date quite quickly, as material and design changes take place on even the most well-designed and managed projects. The main contractor must therefore update the plan during the project and keep it for two years after the project ends. Whilst the duty to produce and use a SWMP no longer exists, many constructors do still use these as a form of good practice.

5.8 Want to know more? Site waste management plan

The UK organisation WRAP has a series of tools and templates that will enable various types of projects and sizes of contractors to take on the site waste management plan (SWMP) way of working, and if undertaken in the spirit that the plans have been intended, reduce waste and keep their activities within all legal frameworks.

http://www.wrap.org.uk/content/site-waste-management-plans-1

A WRAP case study explains how a good practice SWMP was used to plan and monitor waste reduction and recovery on this £35 million higher education college project in Newcastle-Under-Lyme. The project included managing the whole as three projects, each with a separate SWMP developed early on in the planning phase. The plan utilised a site environmental committee that had oversight of all construction disciplines and helped achieve a recycling target of 70%.

http://www.wrap.org.uk/sites/files/wrap/WRAP%20SWMP%20Case%20Study%20BAM.pdf

There are alternatives to waste management, and some have been highlighted by the UK Green Building Council's Sustainable Community Infrastructure Task Group, including waste collection systems, diverting waste away from landfill, and recovering energy from waste.

http://www.ukgbc.org/content/sustainable-community-infrastructure-task-group

5.3.5 Dealing with waste: The three Rs—reduce, reuse, and recycle

Whilst there is the possibility of reducing the quantities of waste produced on construction sites to virtually zero, the preceding waste statistics show that this is currently a distant goal. Three main methods of minimising or effectively dealing with site waste are summed up by the three Rs, shown in Table 5.11.

The methods of reducing, recycling, and reusing materials, although connected, are quite separate when trying to discuss their impact on construction processes and procedures. Recycling involves the identification of recyclable material on site, separation of this material, and transportation to approved recycling centres. Reuse of materials involves the design team and contracting team working together to design for, appropriately identify, and safely use again items of the required quality. The design issues are discussed in Chapter 4. Some of these issues are shared. However, it is apparent that a fourth 'R' is required for a virtuous circle to exist—the use of recycled materials and reused components (Figure 5.6).

5.3.5.1 Reduction in materials coming onto site

Probably the most fundamental and efficient method of reducing waste due to materials coming onto site is limiting the amount of materials and components that have been included in any design. This area can be improved through early involvement of the contractor and the supply chain in the design process. This is discussed in more detail in Chapter 4.

Over-ordering is often used as a method of reducing the risk of delays due to shortages of materials and components. However, beyond the costs associated with materials and components that are not used, there is also

Table 5.11 Methods of minimising or effectively dealing with site waste

The 'R'	The Intention
Reduce	Limit the amount of materials being brought on to a site to those that are needed and no more.
Reuse	Use materials that have come from the same site or elsewhere that are fit for purpose.
Recycle	Take materials and through a collection and sorting process make them available for processing, intending their use for another product or material.

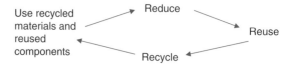

Figure 5.6 Virtuous circle associated with material and component use on construction sites.

the environmental impact and associated costs when they are disposed. The National Specialist Contractor's Council estimates that five landfill sites could be saved every year if the construction industry rid itself of its culture of over-ordering. If a combination of subcontractors, quantity surveyors, and the main contractor can meet to discuss the layout of the site and the materials needed, then the exact quantities required can be more accurately predicted for the project. This in turn should then reduce or remove the need for over-ordering materials. Manufacturing components, both large in the form of complete modules or small in the form of preassembled subcomponents, can, if the system used is well thought through, cut down waste considerably. Off-site 'production line' processes have been proven to decrease waste by optimising cutting patterns and schedules. The use of a permanent or flying-factory facility will allow materials to be delivered directly to the facility, eliminating much of the packaging needed on site and reducing weather damage and handling impacts because of better storage conditions (NSCC 2007).

Good packaging without using over-wrapped materials can also reduce waste. A happy balance can be drawn where enough packaging is used to reduce any possible breakage and good identification whilst utilising reusable packaging (e.g., returnable pallets and crates), particularly for vulnerable and/or high-value materials and products. As mentioned in Chapter 4, making use of standardised components wherever possible (bearing in mind that site tolerances still have to match the expectations of the designed components) can go a long way to eliminating waste from cutting. This strategy or a similar approach, that of ensuring products and materials are cut to size, will eliminate cutting. Storage of these standardised, preassembled, or precut materials will also influence waste. Dry, well-organized, and secure stage areas, designed to allow the use of the appropriate mechanised, lifting, and sorting devices is a must for all well-managed sites and reflects good standards of safety, all values that lead to sustainable construction. 'Just-in-time' approaches can also remove or reduce the need for storage, and this is reflected in many management techniques such as 'lean construction'. The use of lean construction principles (and at this point the author must point out that being 'lean' is often a philosophy as much as a management technique, and the use or misuse of this term can cause much debate) will impact issues such as double handling and will factor out any activities that do not add value to a project. Waste is obviously one of those. Whilst the reader might at this point accuse the author of leading them up a dead-end road, lean construction is a big topic, and one is strongly advised to refer to the following textbox as well as the article by Agyekum and colleagues (2013) for understanding.

5.3.5.2 Reuse

For a design team, client, or contractor, reusing a building with minimal intervention is the optimal solution from a materials point of view. However, if the needs of the client demand a level of building performance, location,

5.9 Want to know more? Lean construction

The principles behind lean construction are similar to those behind its predecessor, lean manufacturing, and they hinge upon trying to deliver to the client whatever was agreed on to be 'right' the first time. The central premise is to eliminate waste, specify values from the client's point of view, identify and use processes that deliver that which the client values, and eliminate all non-value-adding steps, incorporating continuous improvement. In many ways sustainable construction and lean construction have distinct overlaps, many of which are focused on getting more from less, whether it is to the benefit of the client or to the wider community. Reducing waste limits the amount of recycling and reuse that has to be undertaken, and suitable processes that support the value stream, such as good storage and handling (Agyekum *et al.* 2013), can be lean and sustainable.

or space that cannot be met by 'the' or 'an' existing building, reclaiming components is the next best solution. (Chapter 4 discusses the interaction between materials and their reuse or recycling.) This reclamation might involve using existing parts of a building in situ, such as façades or foundations/structural elements, with the obvious proviso that they will be fit for purpose when the demands/imposed loadings of the new structure are operational.

Taking items from other buildings that have become redundant or have been altered entails transporting these items to site but does allow another level of reclamation. This level might entail the complete use of a structural frame or perhaps a floor comprising a number of standard decking units. Smaller features such as windows, doors, and shutters are readily available from organisations such as Salvo, which operates in the UK and the USA (http://www.salvo.co.uk). This organisation allows contractors to source and sell reclaimed components and even complete buildings where relevant. Some components may not be immediately usable, and then a degree of reconditioning may be appropriate before reuse. The ability or degree of reconditioning will depend upon a number of factors. Items of electrical and mechanical plant will have to adhere to regulations governing their safe use. These items would be best refurbished by the original supplier if possible. Often these are no longer in business or are unable to undertake such as task; therefore; more specialist organisations may be able to recondition or refurbish separate or complete units so that they can be used again. At the decision stage within this process, elements such as the efficiency, compatibility, and longevity of the component within the context of whether it will be matched to more modern counterparts needs to be assessed. If the refurbished unit will be functioning appropriately for a long enough timespan, then the decision to use it would be appropriate. Refurbished components may have an advantage in that another refurbishment may be possible, whereas some more modern components may lack the surrounding infrastructure to allow their refurbishment. However, some items on site have value attributed to the materials that they are composed of, but they may not be reclaimed as a complete unit. Recycling or 'recovery of energy' are the next two levels of use.

A more complete hierarchy of steps in waste management is the Delft ladder, which starts at prevention, which is the first priority, followed by a succession of less and less preferable options until the final one, waste designated for landfill. Later systems use a more sophisticated approach where the options available are not linear and allow a range of options rather than one proceeding to the next.

1. Prevention
2. Construction reuse
3. Element reuse
4. Material reuse
5. Immobilisation with useful application
6. Immobilisation
7. Incineration with energy recovery
8. Incineration
9. Landfill

(Hendriks & Dorsthorst 2001)

Reusing materials can be quite different depending on the makeup of the material or component at hand, and some advice concerning several of the more common materials can be seen in Table 5.12. (Additional information from WRAP is indicated at the end of each piece of advice).

Table 5.12 Advice concerning the reuse of several of the more commonly used construction materials

Material	Reuse Advice
Timber	The sources of site-based timber that can be reused or reclaimed are numerous, including boarding, floorboards, timber sections such as rafters, doors, window frames, and fencing. Temporary formwork can typically be reused four times before disposal if suitably looked after. If reusing wood on site, you should always check it first to ensure it is of suitable quality (stress grading etc.) and fit for purpose for the intended use, and store it appropriately. http://www.wrap.org.uk/sites/files/wrap/MDD013%20Final%20Report%2011.03.10.pdf
Bricks, blocks, and tiles	Bricks, blocks, and tiles in good condition can be reused on site for construction if they meet the specification laid down for that particular use. This may include meeting the relevant standards. Reclaimed bricks and tiles in particular can often offer additional character or the chance to use a planning-related requirement where another 'off the shelf' material might not comply. Blocks and bricks that do not offer the correct specification can be used in positions where the weather resistance of structural bearing capacity is not felt to be appropriate. http://www.wrap.org.uk/node/17006

(Continued)

Table 5.12 (Continued)

Material	Reuse Advice
Plasterboard	Most contractors and subcontractors will endeavor to make use of offcuts and leftover plasterboard sheets. However, where the fire resistance or size restrictions make this impossible, a number of organisations are willing to offer these materials to other contractors. As plasterboard, like timber, is vulnerable to the effects of moisture, dry storage is vital. http://www.wrap.org.uk/sites/files/wrap/Cost%20effective%20solutions%20report%20-%20final.pdf http://www.wrap.org.uk/sites/files/wrap/Case%20study%20-%20Plasterboard%20waste%20recovery%20from%20smaller%20building%20sites.pdf
Inert materials	Inert materials either from demolition or from site operations such as concrete, brick, asphalt, soils, and aggregate can be reused on site as hardcore or for backfill. Topsoil can be reused for landscaping or as part of compost once necessary tests have been carried out (see the section in this chapter devoted to soils). http://www.wrap.org.uk/sites/files/wrap/66-OnsiteRecycledAggregates.pdf http://www.wrap.org.uk/sites/files/wrap/Quadrant%203%20Case%20Study%20.pdf
Packaging	Many suppliers will discuss the return of packaging if suitably communicated in advance, and this is probably the best method of reuse. Pallets and crates should always be reused where sustainable construction is valued, rather than letting them go to landfill. Some packaging that was used to protect vulnerable materials can also be used to continue to protect these on site or be used to imaginatively protect finishes, particularly on flooring. http://www.wrap.org.uk/sites/files/wrap/GG606_final.pdf

5.3.5.3 *Recycling on site*

Unsurprisingly when constructing a building or structure, the generation of waste and therefore the associated issues related to recycling are at their greatest when the quantity of trades and amount of construction work are also at their highest. Katz and Baum analysed the waste produced by 10 large sites (7000–32,000 m^2 of built area) constructing residential properties. The sites generated an estimated 2 m^3 of waste per m^2 of built floor area, and from this data the authors developed a model that represents this waste generation (Katz & Baum, 2011).

From this type of data and other sources, it is possible to estimate the quantities of wastes and through further analysis, estimations of the recyclable content. One of the associated issues connected with the proficient recycling of construction materials and waste is the sorting of the materials into usable amounts. Wang and colleagues identified a number of critical success factors that allow effective construction waste sorting on Chinese sites: (1) manpower, (2) market for recycled materials, (3) waste sortability, (4) better management, (5) site space, and (6) equipment for sorting of construction waste (Wang *et al.* 2010).

An overarching organisation, the Construction Recycling Alliance, represents companies within the UK construction industry that have affirmed that they are interested in reducing, reusing, and recycling unwanted material from construction sites. The alliance's main influence is through the pledge that members make to commit to reducing, reusing, and recycling. (http://constructionrecycling.co.uk/). From the NSCC (2007) document *Reduce, Reuse, Recycle: Managing Your Waste*, the author has adapted a five-point plan for use on construction sites:

1. Plan
   (a) Pre-project, identify the types of waste you are likely to generate from the design and specification details.
   (b) As each type of waste is identified, record how much of each is being disposed of, undertaking an audit of each skip and estimating the percentage of each waste type by volume.
   (c) Identify the best options for recycling the different types of waste.
2. Liaise
   (a) Distinguish any specific disposal or recycling requirements related to any unique or general materials.
   (b) Liaise with waste management contractors to find out how they require waste to be segregated and stored.
   (c) Confirm that any requirements stipulated by waste management contractors are both achievable and cost effective.
   (d) If space or other limitations means that waste segregation cannot be undertaken on site, identify a local waste transfer station that will collect and sort your mixed-waste skips.
3. Establish site segregation
   (a) Position different containers for each of the types of waste identified in the site waste plan (if being used), with the agreement of any stipulations of waste contractors.
   (b) Clearly label the containers, preferably with appropriate signage, detailing which material can be disposed of in each one. (Use simple signs such as 'Timber only; no plastics'. For details of a suggested colour-coding system see Table 5.13, 'Colour coding your waste' [NSCC 2012].)
   (c) Position any waste receptacles so as to minimise the distance site workers have to carry materials to dispose of them.
   (d) When the site is spread over a large area, use smaller intermediate bins that can be bulked up into larger segregated skips and ascertain from your waste management contractor the different sizes of container they prefer or supply.
   (e) Secure the containers to prevent cross-contamination from dumping by trades working out of hours or by non-site personnel.
4. Training
   (a) To enable the whole site to effectively use the waste segregation and collection system, training is needed alongside periodic updates focusing on the detail of segregation, waste movement, and ensuring safety.

(b) Periodic monitoring of site waste procedures is needed to ensure compliance.

(c) Some form of reward system either to individual employees or to subcontractors will encourage compliance. Site staff can be motivated by establishing targets and offering bonuses.

5. Appraisal of the system

(a) Site managers should periodically inspect segregated material to ensure that contamination is minimised.

(b) A system can be introduced to elicit feedback from site staff.

(c) Measure and monitor progress and communicate this either on a site-specific website or by more direct pinning up of figures saved to show how much waste has been saved. If appropriate, cost savings or estimates of carbon emission could be displayed.

Table 5.13 Colour coding for different waste materials

Material	Colour
Plasterboard	White
Inert	Grey
Mixed	Black
Wood	Green
Hazardous	Orange
Metal	Dark Blue
Packaging	Brown
Plastics	Purple
Glass	Light Blue

5.10 Want to know more? Waste calculators

A useful tool developed by WRAP in early 2013 as an excel spreadsheet file with advice from the UK Construction Group is the Carbon Calculator. It is intended to calculate the impact of construction and demolition waste in the form of carbon saved when reusing recycling and waste prevention decisions. The tool uses a series of carbon factors in the DEFRA/DECC company reporting factors for material impacts (Annex 14).

Carbon Calculator for Construction and Demolition Waste:
http://www.wrap.org.uk/content/carbon-calculator-construction-and-demolition-waste-0

Whilst energy from waste could be seen as controversial by some commentators who espouse that combustion of waste creates toxic emissions that then need attention, energy from waste that might go to landfill can be an alternative route. Another WRAP spreadsheet tool shows a full list of operational energy from waste plants in the UK. The spreadsheet categorises these plants by technology and waste type.

Energy from waste sites throughout the UK:
http://www.wrap.org.uk/node/15031

focus (such as Sustainability Reporting). There are a large number of these systems, so only a selection of the more commonly used types have been included in the following section.

6.3.2.1 Sustainability reporting

Sustainability reporting comes from the same stable of data declaration as financial reporting. The data would be presented in a report that can be used to assess the sustainability-related performance of a company or organisation. Whilst the linkages with sustainable construction can be relatively loose, there are four key areas: economic, environmental, social, and governance performance, which link to many of the activities of construction-related companies. This method of reporting might also be included in or supplement reporting related to the triple bottom line, or corporate and social responsibility (CSR) (more information related to business behaviour can be found in Chapter 7). A further reporting system that takes all of these non-financial aspects of a company's activities and combines them with more traditional financial information is termed integrated reporting. Integrated reporting is likely to be the major form of reporting in the future when both financial and sustainable sides of an organisation's or company's work need to be seen in context.

One of the main organisations that support sustainability reporting is the Global Reporting Initiative (GRI 2014), which is a not-for-profit organisation that promotes economic, environmental, and social sustainability. Practically, GRI provides a comprehensive sustainability reporting framework that is widely used in many countries and therefore allows for easy comparisons to be made. According to GRI, there are a range of internal and external benefits for organisations that engage in sustainability reporting, and these can include a better understanding of the risks and opportunities. Companies can become more aware of the link between financial and non-financial performance, which can in turn influence long-term management strategy, policy, and business plans. With access to this form of sustainability-related data, processes can be streamlined, costs reduced, and efficiency improved. Depending on access to the data from others, performance can be benchmarked and assessed with respect to competitors, possible collaborators, legislation, and performance standards. This forewarns of issues that might lead to failures in reaching a sustainable solution, and by inference, financial and CSR targets. This, in turn, influences a company's reputation and brand loyalty, demonstrating how the organisation leads and influences expectations about sustainable development.

6.3.3 The natural step

The Natural Step (TNS) framework was developed in Sweden in the 1980s and is a scientifically founded tool that guides the decisions behind sustainable development to solve broad and wide-reaching questions and allow well-informed decisions to be made (TNS 2014a). The framework has been used by many large companies and organisations, such as ARUP, Balfour Beatty, Wilmot Dixon, and Carillion as part of their strategic decision-making processes.

The TNS framework uses four sustainability principles that are intended to allow the transition to a sustainable society (TNS 2014a, reproduced by permission of The Natural Step):

1. Eliminate our contribution to the progressive buildup of substances extracted from Earth's crust (for example, heavy metals and fossil fuels).
2. Eliminate our contribution to the progressive buildup of chemicals and compounds produced by society (for example, dioxins, PCBs, and DDT).
3. Eliminate our contribution to the progressive physical degradation and destruction of nature and natural processes (for example, overharvesting forests and paving over critical wildlife habitat).
4. Eliminate our contribution to conditions that undermine people's capacity to meet their basic human needs (for example, unsafe working conditions and not being paid enough to live on).

Whilst the use of the word *eliminate* could be seen as an unrealistic goal when the development of any building or construction project involves considerable general disturbance, TNS does explain:

> At first reading, the system conditions and basic principles might seem to imply that we must rid society of all materials extracted from the earth and all substances produced by society and that, further, we must never disturb a natural landscape. But that's not what they mean…. It is, rather, that our industrial system has developed so that substances extracted from the earth and produced by society will continue to build up indefinitely in natural systems (TNS 2014a, reproduced by permission of The Natural Step).

An innovation that TNS uses is a method called 'backcasting' where an organisation, or even an individual, can take stock of their current situation and look at a future state the organisation wants to reach. This then enables a detailed look at the journey that needs to be undertaken between the current and future states. This might be undertaken by responding to these three questions:

Does your organisation have a definition of sustainability?
What is, with reference to this definition, the gap in your sustainability profile?
What are you doing, at the strategic level of the organisation, to bridge that gap?

The organisation can then prioritise a series of actions to ensure that the strategy is (1) moving in the right direction (towards sustainability), (2) uses flexible platforms that avoid abortive investments, and (3) makes appropriate business decisions that reflect improvements in sustainability but that also offer an adequate return on investment.

Figure 6.1 shows the organisation's process towards a desired sustainable future (TNS 2014b).

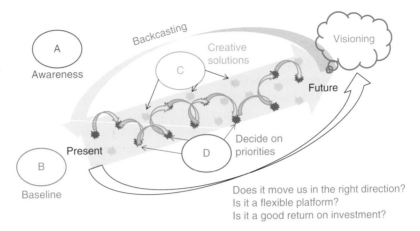

Figure 6.1 The Natural Step ABCD Planning Method (Reproduced by permission of The Natural Step).

As Figure 6.1 shows, the A to B to C to D process runs from an awareness of what needs to be achieved, a baseline mapping, to creating a vision and deciding on priorities for a number of actions. This approach is based on systems thinking, driving organisations/individuals to think, set goals, and develop realistic strategies for moving forward.

6.3.4 ISO 14000 Environmental Management

Much like The Natural Step (which was founded in 1989), environmental management systems also provide an overarching strategic approach to organisational systems, approaches, and actions, to enable appropriate planning. The first formal environmental management system (EMS) developed by the British Standards Institute (BSI) was BS 7750, first published in March 1992 (BSI 1992). Whilst BS 7750 was being further developed, the European Commission was scoping its proposal for an eco-audit scheme,

---

6.1 Want to know more? ISO Standards

ISO Standards are international standards containing the current guidance for a large range of processes and products. They incorporate many more local standards such as the UK's British Standards from the BSI or the German DIN standard.

ISO defines a standard as 'a document that provides requirements, specifications, guidelines, or characteristics that can be used consistently to ensure that materials, products, processes, and services are fit for their purpose' (ISO 2015d).

This form of organised and to a greater extent agreed standardisation can help to harmonise technical specifications of products and services making industry more efficient and breaking down barriers to international trade.

and this was published as the Eco-Management and Audit Scheme (EMAS) (EU 2015), adopted by the Council of Ministers on June 29, 1993.

The standard ISO 14001 was developed because of the need for improved environmental performance expressed at the United Nations Conference on Environment and Development in Rio de Janeiro in 1992 (ISO 2015a). ISO (the International Organisation for Standardization) was charged with creating an internationally recognised environmental management system (Bansal & Bogner 2002).

Two EMS standards are available to organisations in European Countries: the International Standard ISO 14001 and the European EMAS system. The EMAS regulation was revised in April 2001 and was based on ISO 14001. One of the principle benefits of this meant it made it easier for ISO 14001–certified organisations to progress to what is considered the more rigorous EMAS regulation. A further revision of ISO 14001 was released in 2015 (ISO 14001:2015). It is expected that this revision will include stipulations for a high-level structure for management systems and a description of the future challenges for environmental management systems and ISO 14001:2015 (Lloyds Register 2014). (To clarify, the ISO 14001:2004 standard contains the actual requirements that an organisation has to comply with to become certified to the ISO 14000 standard).

Whilst the use of the ISO is more prevalent than EMAS within the construction industry, ISO 14001 is generic in approach and is not tailored to the construction industry's way of working, nor does it lay down any specific benchmarks that a contractor, architectural practice, or specific construction project can follow. ISO 9000, the international standard related to quality management (ISO 2015b), is similar in approach to ISO 14001 and both relate to the process of production, rather than to the product (in this case a building or structure) itself. As with ISO 140001, certification is carried out by third-party organisations rather than being awarded by ISO directly. The ISO 19011 audit standard applies when auditing for both 9000 and 14001 compliance (ISO 2011).

The two parts of the ISO 14000 suite of standards that relate to EMS are ISO 14001:2004, which details the requirements for an EMS, and ISO 14004:2004, which provides more general EMS guidelines. If a construction-related company or practice adheres to the procedures laid down in ISO 14001 it can, first, identify and control the environmental impact of its activities; second, it can continually improve its environmental performance; and finally, it can implement a systematic approach to setting, achieving, and recording environmental objectives and targets. The use of the standard can have advantages for the organisation both internally and externally. Internally, employees or partners can be persuaded that the organisation is indeed taking their environmental responsibilities seriously, thus possibly aiding recruitment of high calibre staff. Externally, customers, clients, and outside agencies can be assured that the organisation will be compliant with the ISO regulations and that the attitude of the organisation is serious in relation to its impact on the environment.

The possible benefits that could be associated with adherence to the ISO 14000 suite of standards include logical advantages to the adoption of a management system, with further improvements associated with a focus on the environment; improved process efficiency through the use of fewer raw materials and less energy; reductions in the generation of waste and any associated disposal costs; and an increased use of salvaged or reusable materials and products.

Other published standards, including ISO 14040 (described in more detail in the next section), can be used in combination with ISO 14001, allowing organisations to pay attention to their local and wider environmental performance. This may include the use of environmental labeling and verification and life cycle assessment (LCA), including the life cycle of their products/services outside the normal boundaries associated with the company's traditional role. The communication of a company's environmental performance is much akin to the previously mentioned sustainability reporting.

Other environmental management tools developed as part of the ISO 14000 series (ISO 2015c) include the following:

ISO 14004, which complements ISO 14001 by providing additional information, explanation, and guidance.

ISO 14031 can guide an organisation to evaluate its environmental performance, addressing the selection of suitable performance indicators. This information can be used as a basis for internal and external reporting on environmental performance.

The **ISO 14020** series of standards is a range of different approaches to environmental labels and declarations, including eco-labels.

ISO 14040 standards give guidelines on the principles and conduct of LCA studies that provide an organisation with information on how to reduce the overall environmental impact of its products and services (more details in the LCA section of this chapter).

ISO 14064 (in three parts) is an international greenhouse gas (GHG) accounting and verification standard, which provides a set of requirements to support GHG emission reduction initiatives.

ISO 14065 complements ISO 14064 by specifying requirements to accredit or recognise organisational bodies that undertake GHG validation or verification using ISO 14064.

ISO 14063 relates to environmental communication. ISO Guide 64 provides guidance for addressing environmental aspects in product standards. Although primarily aimed at standards developers, its guidance is also useful for designers and manufacturers.

6.3.5 Life cycle assessment

LCA systems are not confined to the assessment of buildings or construction projects. Standards exist to guide an assessor as to the procedure and format used when assessing the environmental issues involved with the life cycle of a product.

As described in the section relating to the ISO 14000 series, ISO 14040 (also a British Standard of the same number) (ISO 2006) guides an assessor through a number of stages to analyse the broader picture of the impacts of a product or function from its inception, use, and finally disposal (and in some instances reuse or recycling) of the product's component parts (see the definitions of cradle to cradle, etc., in Table 4.2, Chapter 4). Although this standard does not explicitly deal with buildings, the many products and functions undertaken or used in buildings will be covered. The theory related to life cycle analysis is described in more detail in Chapter 4, which discusses life cycle analysis from a designer's point of view, balancing the approaches that can be used when assessing the merits of one design option or project against another.

The latest version of 14040 is ISO 14040:2006 (ISO 2006) (this replaced earlier versions of ISO 14041 to ISO 14043), which describes the framework used by the standard. This framework is split into a series of phases shown below:

- the life cycle inventory analysis phase,
- the life cycle impact assessment phase, and
- the life cycle interpretation phase: reporting and critical review of the LCA, limitations of the LCA, the relationship between the LCA phases, and conditions for use of value choices and optional elements.

LCA is often used to compare one option against another to establish which of the two is better in life cycle terms. To do this, a reference is needed to allow this comparison to be meaningful. The ISO standard uses a particular function, for example, drying hands, where a paper towel–based system is compared with use of a hot air blower. In this relatively simple case, what is compared is the impact of the use of paper when opting for paper towels against the mass of hot air used to dry the hands with a hot air blower.

Virtually all examples of life cycle analysis will involve a decision that needs to be taken about 'how far down or up the line' of the energy/waste/products/discharges the analysis goes. To try to obtain comparability between projects or functions, a system boundary will need to be applied. This boundary will define where the system's processes or related processes start or finish. This cutoff point needs to be clearly articulated and defined for all those taking part in the analysis. Further details relating to this aspect, along with a system for using LCA for a building production site, is shown in Chapter 4 (Section 4.3).

LCA more directly associated with buildings has been researched in a number of projects including ENSLIC BUILDING (ENSLIC 2015), whose aim was to promote the use of LCA in the design phase of new buildings or major refurbishments.

6.3.6 Carbon footprint of products

A carbon footprint can give an estimate of the total climate change–related impact that is connected to a particular product or activity. The process of assessing a carbon footprint shares some of the methods of

LCA, but unlike LCA techniques, carbon footprinting tends to be focused more specifically on a single material or product. The word *carbon* in this instance also includes the other greenhouses gases that might be associated with a product or material, such as those described in the Materials section of Chapter 4 (methane, CFCs, NOx, and SOx). This is normally termed CO_2e; the *e* stands for 'equivalent'; more details can be found in Chapter 4. Carbon footprinting can, if appropriately used and if comparable with other carbon footprints, have a part to play in reducing greenhouse gas emissions by helping to identify sources. The concept behind carbon footprints identifies the contribution of individual products or materials to the greenhouse effect and thus to climate change. Numerous carbon footprint systems have been developed in recent years. However, initially there was no agreed methodology or documentation to allow for comparative analyses. The recent standard ISO/TS 14067:2013 (Greenhouse gases—Carbon footprint of products—Requirements and guidelines for quantification and communication) (ISO 2013) is very much intended (as its title indicates) to allow a common basis for the estimation of quantities of greenhouse gases related to particular products. The British Standards Institution's latest revision of PAS 2050—Specification for the assessment of the life cycle greenhouse gas emissions of goods and services—was published in 2011 (BSI 2011) (PAS refers to Publicly Available Specification). The revision undertaken by BSI was to update the specification for quantifying the life cycle greenhouse GHG emissions of goods and services in line with the latest technical advances and current experience. This specification provides a consistent method for assessing the life cycle GHG emissions of goods and services that relate to the construction industry.

6.3.7 Key performance indicators

Construction industry key performance indicators (KPIs) are national data sets against which a project or a company can benchmark its performance. The data is collected each year from national surveys covering the construction industry and its clients. KPIs have been used within the construction industry for many years and have traditionally been applied to performance related to quality, finance, and time-related issues. The current set of construction-related KPIs can be accessed via the KPIzone (CCI 2015), which, on a subscription basis, will show KPI graphs covering the following:

- Economic performance indicators (e.g., client satisfaction, predictability, profit)
- Environment performance indicators (e.g., energy use, waste)
- Respect for people (e.g., employee satisfaction, sickness absence)
- Housing (economic)
- Non-housing (economic)
- Infrastructure (economic)

- Mechanical and electrical contractors (e.g., client satisfaction, training)
- Construction consultants (e.g., client satisfaction, training)
- Construction products industry (including customer satisfaction, people and environment issues)

There are now KPIs available that are more closely related to the sustainability of the construction process. Constructing Excellence has produced an Excel spreadsheet that is helpful in establishing monthly targets for water and energy use, waste, and commercial vehicle traffic. The KPIs were linked to compliance with the (now-superceded in England) UK's Code for Sustainable Homes and are established as a tool kit with that code in mind. The tool kit can be obtained from the Constructing Excellence (2009) website.

The spreadsheet has a calculator to help set targets with reference to the industry's average performance, or top and bottom quartile if the contractor requires. The targets are adjustable to tailor the spreadsheet to the specific requirements of the construction project. The tool kit automatically produces graphs in three formats:

monthly resource use against target
cumulative performance against target
benchmark performance against national KPI data

All of these can be used to assist decision making around site-based or more global construction company–related business and processes.

6.3.8 BIM

Building information modelling (BIM) is about sharing data in a structured way, often through a series of integrated information technologies and processes. Whilst being a construction-specific initiative, BIM is classified in this text as a 'broader tool', as it is not intended specifically to assess the sustainability of buildings or processes. However, as one of the main aims of BIM is 'reducing capital cost and the carbon burden from the construction and operation of the built environment by 20%' (UK Government BIM Task Group 2015), the linkages to economic and environmental sustainability are clear. The implications of BIM for sustainable construction are also discussed in Chapter 4, which has a site-based viewpoint and explores some of the changes and benefits that BIM is likely to bring to the role of the site manager and other site operatives.

There are many definitions of BIM, some that have the M representing *management* as well as *modelling* and others that are founded around a particular role or profession within the design, building, and aftercare process. A more general description that is admittedly simplistic but acts as a general working handle, is that BIM is the process of generating and managing data concerning buildings during their life cycle. This description is deceptive, as the task of achieving shared data between such a potentially

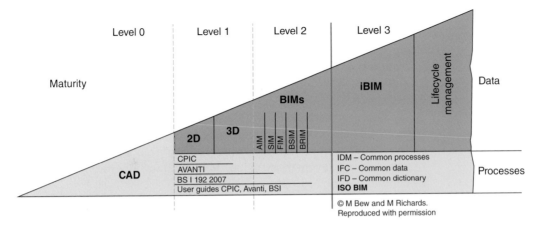

AIM   Architectural Information Model
SIM   Structural Information Model
FIM   Facilities Information Model
BSIM   Building Services Information Model
BRIM   Bridge Information Model

CPIC   Construction Project Information Commitee
IDM   Information Delivery Manual
IFC   Industry Foundation Classes
IFD   International Framework for Dictionaries

Figure 6.2 BIM maturity model (BSRIA 2012). (Reproduced by permission of BSRIA)
Level 0: Unmanaged CAD, probably 2D, with paper (or electronic paper) as the most likely data exchange mechanism.
Level 1: Managed CAD in 2D or 3D format using BS1192:2007 with a collaboration tool providing a common environment, possibly some standard structures and formats. Commercial data managed by stand-alone finance and cost management packages with no integration.
Level 2: Managed 3D environment held in separate discipline BIM tools with attached data. Integration on the basis of proprietary interfaces or bespoke middleware—proprietary BIM. May utilise 4D programme data and 5D elements and feed operational systems.
Level 3: Fully open process and data integration enabled by web services compliant with emerging standards, managed by a collaborative model server. Could be regarded as integrated BIM, potentially employing concurrent engineering processes (Morledge & Smith 2013).

broad-based team of people involved in the inception, design, construction, and aftercare of a building is a difficult one. To help achieve this there are a number of standards that will aid teams to progress their ambitions in relation to BIM. As many of the UK's construction and design companies and consultancies are currently putting plans in place to meet the UK government's requirement for all government-funded contracts to be BIM compliant by 2016, there is a need for an appropriate plan. This plan plots a series of actions and the relevant standards to be used over a number of years to be BIM compliant. Figure 6.2 shows the stages of depth of BIM alongside the timings and standards.

As can be seen from the tools mentioned in Figure 6.2, BIM can vary in use from more simple file-based sharing systems using spreadsheet-type recording methods to 3D CAD with specialist embedded software that links properties of size, time, specification, and so on. There are different levels of BIM compliance from Level 1, which is relatively simple in nature, through Level 2 to Level 3, which is fully integrated. The level that the organisation chooses to work at will influence the degree of effort needed to achieve BIM compliance and also the range of benefits that will be accrued from BIM usage. According to Mott MacDonald https://www.mottmac.com/article/2385/building-information-modelling-bim (reproduced by permission from Mott MacDonald), these benefits potentially include the following:

1. Better outcomes through collaboration
   All project partners—different design disciplines, the customer, contractor, specialists, and suppliers—use a single, shared 3D model, cultivating collaborative working relationships. This ensures everyone is focused on achieving best value, from project inception to eventual decommissioning.
2. Enhanced performance
   BIM makes possible the swift and accurate comparisons of different design options possible, enabling the development of more efficient, cost-effective, and sustainable solutions.
3. Optimised solutions
   Through deployment of new generations of modelling technologies, solutions can be cost-effectively optimised against agreed parameters.
4. Greater predictability
   Projects can be visualised at an early stage, giving owners and operators a clear idea of design intent and allowing them to modify the design to achieve the outcomes they want. In advance of construction, BIM also enables the project team to 'build' the project in a virtual environment, rehearsing complex procedures, optimising temporary works designs, and planning procurement of materials, equipment, and manpower.
5. Faster project delivery
   Time savings of up to 50% can be achieved by agreeing on the design concept early in project development to eliminate late-stage design changes, using standard design elements when practicable, resolving complex construction details before the project goes on site, avoiding clashes, taking advantage of intelligence and automation within the model to check design integrity and estimate quantities, producing fabrication and construction drawings from the model, and using data to control construction equipment.
6. Reduced safety risk
   Crowd behaviour and fire modelling capability enable designs to be optimised for public safety. Asset managers can use the 3D model to enhance operational safety. Contractors can minimise construction risks by reviewing complex details or procedures before going on site.
7. Fits first time
   Integrating multidisciplinary design inputs using a single 3D model allows interface issues to be identified and resolved in advance of construction, eliminating the cost and time impacts of redesign. The model also enables new and existing assets to be integrated seamlessly.
8. Reduced waste
   With the assumption that the data in a BIM model is accurate, exact quantity take-offs mean that materials are not over-ordered. Precise programme

scheduling enables just-in-time delivery of materials and equipment, reducing the potential for damage. Use of BIM for automated fabrication of equipment and components enables more efficient materials handling and waste recovery.

9. Whole-life asset management

   BIM models will act as repositories for information related to products within the building that can assist with commissioning, operation, and maintenance activities.

10. Continual improvement

    Members of the project team can feed back information about the performance of processes and items of equipment, driving improvements on subsequent projects.

Many of these benefits, when achieved, will influence the sustainability of the construction and performance of buildings in a wide range of ways. More information concerning the operation and potential benefits of BIM can be found on a number of country-specific websites such as that of the UK Government BIM Task Group (2015).

6.4   Building- and project-specific tools

Tools that assess a particular aspect of building or process performance are distinct from the broader tools and systems described in the first section of this chapter, as their applicability beyond that of buildings can be limited. They can range from physical measurements taken in buildings that are already built to the predictive software that assesses potential performance from a design. However, they all have one element in common, that they can help the building professional to make judgments concerning the past, current, and future performance of buildings and apply this to present and future decisions.

The numbers and types of these tools vary month by month. Some tools are linked to legislation, and when the statute changes so does the tool that helps the building profession ensure compliance. Technologies improve and a tool focused on a past technology is no longer valid once that technology is superseded. Social impacts on the design and use of buildings often mean that a previous tool, based on a series of assumptions, cannot adequately describe the new conditions created by the changes in the way we live, work, and relax. New facts (or as close to the current facts as we can get) may influence the direction of a tool's focus, such as the reduction of emphasis on current CFC emissions (as their legal use was phased out) and greenhouse gases (their legal use was phased out, although regulations still allow for past emissions due to their long atmospheric lifetime). This all builds to act as a health warning concerning the applicability of tools in the future. Some, such as the monitoring equipment, will have a purpose for many years to come, but their possible integration into existing devices rather than as a standalone network is likely.

Other tools, if used after they are generally deemed obsolete, might not ensure compliance of a code or law, which will leave the operator open to challenge. Therefore, whilst the following descriptions of tools and the accompanying examples should be very informative and were correct as the book was going to press, care should be taken and checks made before use.

6.4.1 Software and paper-based assessment tools focused on assessing some aspect of the sustainability of buildings and structures

Table 6.1 displays the major assessment systems that are commonly used in different countries with an appropriate link to their main web-based resource.

6.4.2 Code for sustainable homes

Domestic and commercial/industrial categories of buildings are obviously used for a range of different purposes but also often have more fundamental differences in their construction and services that warrant different criteria for assessment. Some assessment systems are bespoke to different building types whilst others have separated different themed strands of assessment for more narrow uses, such as schools or shops. There are convincing arguments for these two methods of operation, including the practical simplicities of operation for more general assessment or the level of detail that the bespoke assessments can achieve. Other reasons for the variances between assessment systems can sometimes be accounted for by the genealogy of the system and the characteristics fed forward to the next system. As with exam grading systems, knowing what a 'C' grade means, in terms of comparability with other grades, is key to their currency, but this becomes confused if this alphabetical system changes to a numerical assessment such as 55/100. One of the systems of assessment that has grown from a previous system is the UK's Code for Sustainable Homes.

The Code for Sustainable Homes was the UK's national standard relating to the sustainable design and construction of new homes and for a time has been seriously considered for integration into the UK building regulations (Mark 2014) to reduce the number of assessments required. It aimed to reduce carbon emissions and promote higher standards of sustainable design above the current minimum standards set out by the UK building regulations but, if integrated, would provide a potential upward regulatory push on efficiency of elements such as energy and water consumption. The Code for Sustainable Homes was based upon a previously used assessment system, the Ecohomes assessment method (Ecohomes 2012) used by the UK's Building Research Establishment (BRE). Because of the code's lineage and the focus on homes, the code's applicability is directed to use on

Table 6.1 A selection of the assessment tools and systems that are directly applicable to the built environment

Assessment Method Acronym	Description of Assessment Method	Country of Origin	Notes
BASIX	Building Sustainability Index	Australia, New South Wales	Introduced by the NSW Government, BASIX is a Building Sustainability Index; sets energy and water reduction targets for domestic units. https://www.basix.nsw.gov.au/information/about.jsp
BEPAC	Building Environmental Performance Assessment Criteria	Canada, British Columbia	Developed by the University of British Columbia in 1993.
BREEAM	Building Research Establishment Environment Assessment Method	UK	Developed in 1990, BREEAM is one of the systems with the longest track record. BREEAM covers a range of building types. The method relies on credits being awarded for each assessment area that contribute to a total score. The number of total credits allow the building to receive a 'Pass', 'Good', 'Very Good', 'Excellent', or 'Outstanding' rating. Updated in 2014. http://www.breeam.org/
CASBEE	Comprehensive Assessment System for Built Environment Efficiency	Japan, Japanese Green Build Council (JaGBC)/ Japanese Sustainable Building Consortium (JSBC).	The CASBEE assessment tools are CASBEE for Predesign, CASBEE for New Construction, CASBEE for Existing Building, and CASBEE for Renovation, to serve at each stage of the design process. http://www.ibec.or.jp/CASBEE/english/index.htm
CFSH	Code For Sustainable Homes	UK, governmental system in conjunction with the BRE	Although not now applied to England, the Code for Sustainable Homes is an (applicable in Scotland) environmental assessment method for rating and certifying the design and construction related performance of new homes. http://www.planningportal.gov.uk/uploads/code_for_sustainable_homes_techguide.pdf
Eco-Quantum	Eco-Quantum	Netherlands, developed by the University of Amsterdam distributed through IVAM research	Construction professionals can submit relevant information on environmental matters, which the tool can use to undertake a Life Cycle Analysis related to building products with compliance to ISO 8006. http://www.ivam.uva.nl/index.php?id=373; in Dutch, but Google can translate.
SBTool (previously GBTool)	Green Building Challenge	Originally from Canada, but further developed with help of experts from a range of countries	SBTool as a rating system for certification of buildings requires benchmarks to suit local conditions and has influenced national systems being used in Austria, Spain, Portugal, Japan, and Korea http://www.iisbe.org/sbmethod-2010.

(Continued)

Table 6.1 (Continued)

Assessment Method Acronym	Description of Assessment Method	Country of Origin	Notes
Green Star	Green Star	Australia	Green Star is a system undertaken in conjunction with the Green Building Council Australia, a national, voluntary system that evaluates the environmental design and construction of buildings. Green Star uses environmental measurement criteria with particular relevance to the Australian marketplace and environmental context. http://www.gbca.org.au/green-star/
HK BEAMPlus	Building Environmental Assessment Method	Hong Kong	Supported by the BEAM society, HK BEAM provides a common set of performance standards that can be used by developers, designers, architects, engineers, contractors, and operators. https://www.hkgbc.org.hk/eng/BEAMPlus_NBEB.aspx
LEED	Leadership in Energy and Environmental Design	US	LEED was developed by the US Green Building Council (USGBC). LEED allows building owners and operators to identify and implement practical and measurable green building design, construction, operations, and maintenance solutions. http://www.usgbc.org/leed
NABERS	National Australian Built Environment Rating System	Australia	A performance-based rating system for existing buildings comparing impacts measured from one building against selected comparators. http://www.nabers.com.au
SBAT	Sustainable Building Assessment Tool	South Africa	SBAT was developed by CSIR (the South African Council for Scientific and Industrial Research) for the Sustainable Building Best Practice Awards in 2004 and uses an Excel format in 'lite' version. SBAT is influenced by international best practice and has been altered through use in South Africa to reflect the local context and policy. http://www.csir.co.za/Built_environment/Architectural_sciences/sbat.html
SPeAR	Sustainable Project Appraisal Routine	UK/international	SPeAR was developed by consultants ARUP as a sustainability decision-making tool. One of the more unique features of SPeAR is that it communicates outcomes visually. An integrated set of indicators covers a range of issues based on global sustainability standards. http://www.arup.com/Projects/SPeAR.aspx

a domestic scale rather than on more commercial projects; thus, large blocks of multioccupancy buildings sit less happily within the remit of the code. This integration has now come to pass for use in England, but the code is still used in Scotland, and a number of other UK companies still refer to the levels as a benchmark for performance. For this reason it is appropriate to look into some of the detail of the code.

The current code aims, through the use of six levels of performance and a range of mandatory and optional categories and subcategories, to assess a building for a range of different issues:

energy/CO_2
water
materials
surface water runoff (flooding and flood prevention)
waste
pollution
health and well-being
management
ecology

Homes are/were assessed at the design stage and also postconstruction, with mandatory elements in a range of areas. In particular, it places demands on the designer and contractor to achieve low CO_2 emissions and water consumption. All homes in England must be issued with a code certificate, but currently there is no requirement for the home to be assessed at a particular level for compliance beyond the building regulations. A 'nil-rating' certificate can be issued if a rating has not been undertaken under the auspice of the code. The only homes that do/did need to reach a specific code rating are those built using public funding or those homes built on publicly owned land. These homes needed to achieve a rating of at least level 3, which will involve improvements over and above the standards that are stipulated for current UK building regulations but do not stretch to the more onerous requirements of levels 4, 5 and 6.

Code levels 3 and 4 are generally seen as more achievable in that they may require some renewable technologies to attain compliance but not the scope or quantity of solar, wind, biomass, or other similar technologies needed to reach code levels 5 and 6. Table 6.2 shows the way the code weights and allocates credits to measurement themes. Table 6.3 shows the point scores needed for the higher levels of the code.

More information concerning the UK Code for Sustainable Homes can be found at https://www.gov.uk/government/publications/code-for-sustainable-homes-technical-guidance (UK Government 2010), and other guidance at https://www.gov.uk/government/policies/improving-the-energy-efficiency-of-buildings-and-using-planning-to-protect-the-environment/supporting-pages/code-for-sustainable-homes (UK Government 2015).

Table 6.2 The allocation of credits and weightings within the previously used Code of Sustainable Homes (CLG 2010)

Category	Environmental Subcategories	Credits	Weighting (% contr. to score)
Energy **Contains mandatory elements (CME)**	Dwelling Emission Rate (DER), building fabric, internal and external lighting, low or zero carbon (LZC) technologies, energy labelled appliances, home working space and cycle storage	29	36.4
Materials **CME**	The impact and sourcing of structural and finishing materials.	24	7.2
Water use **CME**	Use inside of exterior water	6	9
Surface water drainage and run off **CME**	Management of surface water run off from the development as a whole and the risk of flooding	4	2.2
Waste, both construction and waste in use. **CME**	Construction site waste management, storage provision for household wastes and composting facilities	7	6.4
Pollution	Global warming potential of specified materials (insulents and refrigerants) and nitrogen/oxygen compound emissions (NO_x)	4	2.8
Interior environment	Use of daylight, insulation against noise and the allocation of private space	12	14
Site management	The provision of a home user guide, membership of the considerate constructors scheme, various construction site impacts and security	9	10
Ecological issues	Ecological value and enhancement of the site including the protection of ecological features; the building footprint	9	12

Table 6.3 The total point scores needed for Code for Sustainable Homes Levels 3–6, (CLG 2010)

Category	Code Level 3 Points	Code Level 4 Points	Code Level 5 Points	Code Level 6 Points
Energy	6.28	10	17.6	18.9
Building fabric	0	0	0	2.5
Indoor water use	4.5	4.5	7.5	7.5
Lifetime homes	0	0	0	4.7
Mandatory points	10.8	14.5	25	33.5
Optional points required for level rating	46	53.4	58.9	56.5
Total points	57	68	84	90

6.4.3 BREEAM

The UK's BRE first published BREEAM (Building Research Establishments' Environmental Assessment Method) in 1990, focusing on offices. This followed a period of research in the late 1980s. This was the first system that took a relatively holistic view of the environmental performance of buildings. Previous assessment systems did exist but tended to focus on one part of the environmental elements that make up the possible impacts of a building. As BREEAM was one of the first systems in use, many of the succeeding assessment methods followed the good practice laid down by BREEAM and similar assessment methods are used in different parts of the world, with alterations that enable the system to be country/region/project specific.

Over the years BREEAM has grown in scope to cover different building types, for example, schools, offices, commercial properties, and so on. A number of the case studies held by the UK's BRE are shown in Table 6.4. More examples can be found at http://www.breeam.org/case-studies.jsp (BRE 2015).

Some major revisions were introduced into the 2008 BREEAM series of assessments. Compared to the previous version, BREEAM 2006, the 2008 versions included a new category of Outstanding, beyond the certified grades of Pass, Good, Very Good, and Excellent. Submetering of certain utilities/consumptions became a mandatory minimum requirement in some areas. The ability to achieve some credits was made more difficult, often through changes to weightings and, importantly, the introduction of a post-construction stage (to check that features assessed at the design stage had been maintained during construction and initial occupation). This final aspect should prove a very interesting exercise as in some cases it has been mooted that buildings sporting a BREEAM Excellent grading have not fulfilled their promise in terms of the performance of the building.

6.4.3.1 BREEAM 2011

BREEAM 2011 (BRE 2011) was expanded in scope to cater for a number of other building types (which under older versions of BREEAM required BREEAM Bespoke Criteria). In addition to the common building types, BREEAM New Construction 2011 could be used to assess the following buildings: residential institutions (hotels, hostels, training centres), non-residential institutions (community buildings, museums, libraries), assembly and leisure buildings (cinemas, information centres, sport centres), and transport hubs (train/bus stations). The updates contained within BREEAM 2011 are needed (as are those in its predecessor BREEAM 2014), as there are a number of new regulations, both advisory and legislative, that need to be reflected within BREEAM's assessment framework. The latest UK building regulations, Part L 2013, (which came into effect on 6 April 2014) (DCLG 2013) function to conserve fuel and power and are intended to drive considerable reductions in energy use in buildings. In addition, new approaches to assessing how much energy/carbon is being used or emitted from buildings are also included. Other influential industry-related metrics include the use of corporate reporting systems that will require in-use data

Table 6.4 Examples of BREEAM projects (Descriptions reproduced by permission of BRE)

BREEAM Version	Building Type	Brief Description	BREEAM Rating
Communitites	Sheffield Housing Company, UK	A joint venture across three sites between Sheffield City Council, Keepmoat Homes, and Great Places. 2,300 new homes are planned for sale and rent on 60 hectares of land over the next 15 years in Sheffield, UK. The programme includes 20 brownfield sites within existing urban areas.	All three projects scored 'Very Good'
Courts	Port Talbot Justice Centre houses	The BREEAM 'Excellent'-rated Port Talbot Justice Centre houses attained an EPC rating of 27. The design of the building regulates its environment through the use of a superinsulated and airtight façade, north-south orientation to minimise heat loss, and solar gain and enhanced daylighting. The building uses grey water recycling and roof-mounted solar panels for hot water to reduce energy and water consumption. The fabric combines the use of recycled materials with modularisation, creating less waste. The building is flexible and can function solely as an office if required.	Rating: Excellent
Data centres	British Geological Survey Data Centre, Nottingham, UK	The building is a data centre in which data processing facilities were rehoused to include a direct replacement for the existing computer suite; the building also provides space for future expansion of computing facilities. The new building is intended to improve the efficiency of the technical facilities in terms of performance, energy efficiency, and space standards.	Rating: Excellent Score: 78.98%
Education	Arcadia nursery, Edinburgh, UK	The Arcadia nursery, Edinburgh, is a new building of approximately 843 m² over two floors accommodating up to 113 children. This project attained 100% in both the Materials and Pollution BREEAM categories. This was achieved by: • selecting materials that scored very highly in the Green Guide, including a large amount of responsibly sourced timber, • connecting to the existing central combined heat and power (CHP) unit on site, • using no refrigerants in the building, • building on a site with low risk of flooding—but also incorporating sufficient sustainable drainage systems (SUDS) to achieve maximum credits. The project also scored very highly in the Management (82%), Health and Well-being (80%), Energy (69%), Transport (86%), Water (78%), and Waste (83%) sections.	Rating: Excellent Score: 82.2%

| | | Rating: Excellent
Score: 74.3% |

Healthcare

Dumfries and Galloway Acute Mental Health Unit, UK

Dumfries and Galloway Acute Mental Health Unit, now known as Midpark Hospital, was opened in 2012 and has six wards and 85 beds.

The major environmental features include the following:

- The prefabricated structure played a major part in reducing the construction waste produced on the project.
- A unique anti-ligature window included in the naturally ventilated windows—specified and installed by Laing O'Rourke—allowed all patient areas to be naturally ventilated, with wind catchers installed for internal areas, in order to reduce carbon emissions.
- High levels of daylighting and views out into the gardens and landscape help to optimise the health and well-being of all patients and staff. There is evidence that the quality and quantity of light have major impacts on human body. In a healthcare facility, patients, visitors, and staff are mostly exposed to artificial light.
- Low-water-use appliances and energy and water submetering have been installed.
- A Low Carbon District heating system uses biomass to provide heating and hot water via the in-site buffer vessels fed by the district heating heat exchangers in the plant room.

Industrial

Diageo Guinness Brewhouse No. 4, St. James Gate, Dublin

Brewhouse No. 4 takes the form of a main two-storey block that houses the brewing tanks and vessels, an extensive ground floor process area, together with upper mezzanine areas giving access to the higher sections of the large brewing vessels. This element of the building rises to a height of about 14 m to parapet level.

The Brewhouse energy usage and demand were designed, modelled, and adjusted in order to ensure a carbon negative building. This has been best achieved through the promotion of energy-use reduction by utilising the process water and innovative hydronics (water as heat transfer medium in heating and cooling). The energy profile has been delivered by the use of the following design features:

- Installation of energy recovery system in the new brew house, which reduces the requirement for steam heating in the brewing process to the building operation, maximises the thermal energy recovered, and minimises vapour losses from the brew house and atmosphere
- Installation of variable speed drives on the majority of process drives
- Installation of a hybrid refrigeration system aiming to use 0°C instead of −4°C
- Improvisation of improved metering throughout the brewing process
- Optimum building orientation design
- Solar shading, and the use of high-specification fabric achieving U-values significantly better than the current Irish Building Regulations
- Enhanced airtightness

Table 6.4 (Continued)

BREEAM Version	Building Type	Brief Description	BREEAM Rating
		• Automatic lighting controls complete with high-efficiency lighting and occupancy sensors throughout the building • HVAC systems that mostly condition through mechanical ventilation with selected fan specific power levels that ensures only a small portion of the building is air-conditioned • HVAC plant has been specified and procured based on low-energy usage, fan speeds, and high efficiency. • A CHP plant provides electrical energy for the entire building HVAC and process plant and utilises 100% of the thermal energy for building services and process heating.	
Entertainment & leisure	Works 2/Y Ffwrmes, Llanelli, UK	Y Ffwrnes is a theatre complex that will provide a venue for drama, musical theatre, opera, concerts, dance, cabaret, public meetings, and social functions. This is the first theatre in Wales to have full accessibility, allowing wheelchair access to the fly-tower, allowing disabled people the opportunity to be trained and gain employment as lighting and backstage technicians.	Rating: Excellent (utilising the BREEAM 2008 UK Bespoke route)
Offices	Washington Plaza, Paris	Washington Plaza is an office building of 47,097 m² gross lettable area, located at 42 rue Washington 75008 Paris, off the Champs-Elysées. Washington Plaza excelled in BREEAM categories such as Health and Well-being, Energy, and Pollution. Green spaces, lounge-style spaces, and work spaces were designed to improve access, use, and circulation of people, thus creating alternative areas for socialising. A monitoring system has been deployed to provide precise and detailed measurements of the electrical consumption. This equipment is a true innovation because it enables the breakdown of consumption by use (IT, lighting, heating/cooling). Green spaces were created in courtyards that were initially very sparse in that respect.	Rating: The building has been certified BREEAM In-Use International with the following ratings: Part 1, Asset Rating: Very Good, 67% Part 2, Building Management: Outstanding, 88%
Other buildings & mixed use developments	De Balk van Beel, Leuven, Belgium	The former site of the InBev brewery is being redeveloped into a car-free district with 1,200 low-energy apartments and 70% open and public spaces, in which 5,000 people will be able to live and work. The development's aspirations are being delivered by the use of the following design features: • decentralised generation—local distribution—local supply: matching supply and demand for heat and power • demand for heat reduced through insulation and airtightness	Rating: Outstanding Score: 87.81%

Prisons	Weald Wing, HMP Maidstone, UK	• low-energy housing: K-level 26 W/m^{2}K (BAU = K45) • maximisation of renewable energy sources • local CHP designed to match heat demand in the district • organic gas through local fermentation via intercommunal waste collector The Ministry of Justice, National Offender Management Service required the major refurbishment of the Weald Wing accommodation block in HMP Maidstone due to the aging nature of the facilities. The Weald Wing is a four-storey Victorian structure with seven listed features and houses occupants at the category C listed prison.	Rating: Excellent Score: 71.01%
Residential	Carnegie Village, Leeds, UK	A student residential development within the existing Headingley Campus of Leeds Metropolitan University, providing 479 study bedrooms in 'cluster flat' and 'townhouse' arrangements. The commitment of Leeds Metropolitan University and the University Partnership Programme to environmental issues required this student accommodation to focus on sustainability and improving environmental performance.	Rating: Excellent Score: 76.10%
Retail	Fornebu S, Norway	Fornebu S is a shopping mall located in a combined commercial and residential area just outside of Oslo. This extensive development is on a brownfield site, home of the largest airport in Norway until 1998, and is predicted to be as large as a medium-sized Norwegian city by 2020. The building is owned and run by KLP Eiendom and houses more than 80 shops and restaurants, with a total area of 65,000 m^2, including parking. It is the first shopping mall to receive a BREEAM Outstanding certification. The most important features associated with the building's energy performance are: • an airtight and well-insulated building envelope of which only 7% is windows/doors • effective heat recovery from the technical systems • district heating and cooling using seawater for heat exchange • zoning and sensors to optimize ventilation • LED lights in both shops and common areas • 1043 m^2 solar cells producing 140,000 kWh/year • energy-efficient and water-preserving sanitary equipment • comprehensive metering and an energy monitoring system to help optimise energy and water use The calculated amount of energy consumed in the building is 85.9 kWh/m^2 (according to NS 3701 [the appropriate Norwegian energy standard]), which is about a 60% reduction against the Norwegian regulatory level.	Rating: Outstanding Score: 89.5%

Source: BRE 2015.

allowing comparisons of wider building performance. In addition, changes in the market for more environmentally appropriate buildings should support the evolving knowledge base of data relating to how these buildings actually perform when in use, and allow clients to make use of this data when valuing buildings, whether for rental or for outright sale.

BREEAM 2011 is a subtle but fundamental update of the previous versions. One of the most noticeable changes is that the different versions tailored for different building types and uses were previously all separate but have now been consolidated into one technical guide, simplifying the process of locating advice and guidance. The biggest changes that BREEAM 2011 introduced over the previous assessment criteria are contained within the management category.

6.4.3.2   BREEAM UK new construction 2014

The latest update of BREEAM came into force in May 2014 and is focused on new construction. The alterations and updates from the 2011 version are not fundamental. The alterations improve, clarify, and evolve existing criteria to match the changes in regulations, standards, and best practice. The technical guide uses 51 assessment issues arranged over 10 sections that define the scope and give guidance on how to measure a scheme's match (BRE 2014). It improves linkages to other schemes such as BREEAM

Table 6.5 BREEAM new construction non-domestic category weightings for fully fitted out (BRE 2014, reproduced by permission of BRE)

Category/Issue?	Percent Weighting		
	Fully Fitted Out	Shell Only	Shell and Core Only
Management	12	12.5	11
Health and well-being	15	10	10.5
Energy	15	14.5	15
Transport	9	11.5	10
Water	7	4	7.5
Materials	13.5	17.5	14.5
Waste	8.5	11	9.5
Land use and ecology	10	13	11
Pollution	10	6	11
Total	100	100	100
Innovation (additional)	10	10	10

Table 6.6 The number of credits available for the management category (BRE 2014, reproduced by permission of BRE)

Description	'MAN' Category	Credits
Project brief and design	1	4
Life cycle cost and service life planning	2	4
Responsible construction practices	3	6
Commissioning and handover	4	4
Aftercare	5	3

communities and improves process documentation to help assessors and those who are party to the assessment (Yetunde 2014) (Tables 6.5, 6.6).

6.4.3.3   Hong Kong BEAM and BEAM Plus

The Hong Kong building assessment system was originally based on the UK's BREEAM system but has subsequently evolved to take local issues on board to change various parts of the assessments. This reflects the many multistorey apartment blocks in Hong Kong, along with differing climate-related factors, such as the need for comfort cooling rather than heating many times during the year. HKBEAM is now promoted and developed by the BEAM Society Limited, a non-profit organisation (BEAM 2015a). http://www.beamsociety.org.hk/en_index.php

BEAM Plus Version 1.2 for both new and existing buildings was officially launched on 3 July 2012 and became mandatory on 1 January 2013. Version 2 of BEAM Plus was officially launched on 25 September 2015 (Beam 2015b). Other versions include BEAM 4/04, 5/04, and BEAM Plus Interiors, which are also available on the BEAM website (http://www.beamsociety.org.hk) (BEAM 2015b).

6.4.4   LEED

Leadership in Energy & Environmental Design (LEED) is a building certification system that was developed by the United States Green Building Council (USGBC 2015) and was launched between 1998 and 2000. The major focus of LEED is the following:

energy savings
water efficiency
reductions in CO_2 emissions
improved indoor environmental quality
stewardship of resources and sensitivity to their impacts

LEED, in a similar fashion to BREEAM, offers a suite of different types of assessments to allow the appropriate assessment of different buildings and uses. However, unlike BREEAM, LEED does not use trained accredited professionals (AP) to assess the components but does offer an extra credit if a trained AP is used. Each of the members of the suite uses a common series of themes to assess the building or project. These areas of measured performance include the following:

Sustainable site
Water efficiency
Energy and atmosphere
Materials and resources
Indoor environmental quality
Locations and Linkages
Awareness and education
Innovation in design

Regional priority credits (which differ depending on the US state that the project is being built in)

Current (at time of writing) LEED rating systems:

New Construction & Major Renovations
Existing Buildings: Operations & Maintenance
Commercial Interiors
Core & Shell
Schools
Retail
Healthcare
Homes
Neighbourhood Development

One of the most popular LEED versions is New Construction & Major Renovations (http://www.usgbc.org/resources/list/all/new-construction). This LEED standard, LEED (for New Construction) v1.0, was released in 2000 as the first LEED rating system aimed at new commercial office buildings. In the standard's present form, LEED for New Construction can be used in relation to a wide range of building types, including libraries, hotels, offices, churches, and government buildings. LEED for New Construction addresses design and construction for new buildings and also assesses larger renovations of existing buildings. Major renovations of buildings that qualify will often include large-scale alterations of building services involving substantial improvements in HVAC systems, alongside the more often encountered envelope modifications and major interior refurbishment. This standard's ability to assess refurbishments currently gives the LEED system an advantage over some other standards, but many of those are likely to respond, encompassing this element in their series of categories in the near future.

A further evolution of the LEED system, Version 4, is now available (USGBC 2015) and brings together a number of suggested adjustments. The USGBC has a comprehensive website that contains updates on the current position of the latest LEED standards as well as a number of training and educational tools and information areas (http://www.usgbc.org).

6.4.5 CEEQUAL

CEEQUAL is a civil engineering sustainability assessment system, introduced in 2003. CEEQUAL deals with similar issues as do building-related assessment systems but has categories that fit large and small infrastructure projects better. CEEQUAL takes a broad view of the extent of the reach of civil engineering as it can assess most aspects of infrastructure work, as well as landscaping and the space between buildings.

CEEQUAL uses two main categories of projects:

1. CEEQUAL for Projects
2. CEEQUAL Term Contracts for use on contracts for maintenance and multiple small works

Table 6.7 CEEQUAL assessment categories and weightings

CEEQUAL Assessment Section	% Weighting
Project Management	10.9
Land Use	7.9
Landscape	7.4
Ecology and Biodiversity	8.8
Historic Environment	6.7
Water Resources and the Water Environment	8.5
Energy and Carbon	9.5
Use of Materials	9.4
Waste Management	8.4
Transport	8.1
Effects on Neighbours	7.0
Relations with the Local Community and other Stakeholders	7.4

CEEQUAL splits the areas of assessment into twelve categories that are weighted according to the percentages shown in Table 6.7. As the table shows, the variety of categories broadly cover project-related issues such as materials and water use but have another facet concerned with looking after those people who are either part of or impacted by the project.

Manuals for each of the above CEEQUAL assessment categories can be downloaded at http://www.ceequal.com/frm_manualdownload.php (CEEQUAL 2015). The final grading varies from a pass, at more than 25%; through good, more than 40%; very good, more than 60%; and finally, more than 75%, which is graded excellent. The actual score appears on the certificate.

A fee is charged to cover the cost of the CEEQUAL-appointed verifier (but not the time of the assessor) and also to cover CEEQUAL's administration costs, the cost of arranging the award presentation at the end of the assessment, and the progressive development of the CEEQUAL scheme. The fee is based on the contract value of the project or contracted works or, if applying early in the process, on the client's or engineer's estimate. The assessment fee structure is influenced by the size of the project being assessed and larger projects, with their larger work-related content, cost more to be put through the system. The assessor used is normally one of the personnel of the assessed project who has been through training. Final rating is overseen by an external verifier to ensure uniformity of grading.

6.4.6 The SKA tool

SKA is a simple abbreviation of Skansen, the developer who set up a research project with the Royal Institution of Chartered Surveyors (RICS) and consultant AECOM to measure the environmental effects of an office fit-out in 2005 (RICS 2015). It is an assessment method that has been produced alongside a leading building services and consultancy company. It is seen to be a competitor to other assessment systems such as BREEAM.

Currently, the SKA assesses retail offices and restaurant areas of commercial building activity. The system uses three grades—gold, silver, and bronze—to identify levels of attainment. SKA is designed to be an open, transparent, and flexible tool associated with reasonable costs limiting the scope of the tools to fit-out a building. This inevitably means that no land or locational issues can be assessed. The features that the tool includes are as follows:

- All the design team can view the tool.
- Good practice measures are not credited.
- There is less paperwork than with many other systems.
- The RICS has formally adopted SKA ratings since 2009 (latest upgrade at time of writing 2011), and the original inception was a joint element between Skanska and AECOM in 2005.
- Most projects obtain the silver level of attainment currently; others aim for gold but sometimes pare back some of the more expensive elements and end up with silver.

One of the major advantages of using SKA is the ability of personnel from the company to use the tool to look at a number of scenarios. This affords the consideration of different approaches to a fit-out. This in turn allows a much more flexible and constructive view of assessing the overall performance of a project. Some other systems require an outside assessor who may well be sticking more rigidly to frameworks that have been agreed on centrally. However, the question remains: how can the quality of an assessment be relied upon if such built-in flexibility is a founding feature of the system? To reduce the possible number of less-reliable assessments, a certain number of the assessments are scrutinised officially to ensure quality control. It is a difficult task to balance the flexibility and innovation of any assessment system alongside the more rigorous outside viewpoint related to how 'well' the building is doing. Time will tell whether this more relaxed approach will be taken up by more construction professionals and whether the outcomes will be valued by clients and occupiers or not.

6.5   Post-occupancy assessments

The physical processes that together comprise the commissioning and post-occupancy stages of any construction project aim to aid how a building functions or is used, after handover. These are a fundamental part of ensuring optimum building performance and need to be carried out whether using an assessment system or not. However, there are a number of assessment methods and systems that can aid and inform both of these processes, improving their effectiveness and allowing comparisons between the buildings in question and best practice.

More holistic assessment systems normally refer to the building's performance as a whole. In some situations, such as when the 'soft landings' system (see Section 6.5.2) is being adhered to, assessments require inputs from most parties connected to the design, construction, and post-occupation stages. The less holistic and more focused elements of commissioning are usually more technical and/or directly occupant focused in nature, and these normally lie firmly within the domain of specialist consultants. For this reason only the more general technical methods, such as thermography and air leakage testing, are described in the next section.

6.5.1 Traditional post-occupancy evaluation

Just as we would physically monitor a building concerning its performance (as described in Section 6.6), it would be appropriate to ask the building's occupants and users if they felt the building or structure was working for them. Issues related to water and energy use are connected to occupant and systems behaviour. Some areas of occupant well-being are not necessarily tied to an aspect of a building's performance that can be directly illuminated by looking at a utility bill. Post-occupancy evaluation (POE) is used to assess a building's wider performance, appraising both the numerical and human aspects of how well a building works over a period of time.

What the traditional POE process entails, the time it will take, and the depth at which it will be undertaken, will vary according to the various purposes of the evaluation. POE can be considered a structured assessment of the performance of buildings and facilities. Traditionally, the basis of POE is the ability to systematically review and then evaluate the performance of an occupied building or facility. Finally, the interior and exterior environments in and around the building in question would be adjusted to meet the needs of the occupants, demands of operators, and predicted requirements of best practice. The main advantages of traditional POE are the following:

- It links the perceptions of occupants to the performance of a building.
- It allows the effects of design, construction, and operational strategies to be feedback.
- It forms a sound basis for constant improvement.

Post-occupancy evaluations have a history of approximately 25 years (Cooper *et al.* 1991); Table 6.8 below lists some of the more important publications that have contributed to the development of POE in the earlier years.

Some of the more important publications that have contributed to the development of POE (Bordass & Leaman 2015).

The final document listed in Table 6.8 relates to a research project entitled PROBE (Post-Occupancy Review of Buildings and their Engineering), which ran from 1995 to 2002 funded under the Partners in Innovation scheme (Bordass & Leaman 2002). It was carried out by a range of organisations including Building Use Studies, William Bordass Associates, Energy for Sustainable Development, and Target Energy Services. Further details about PROBE case studies can be located at the Usable Buildings website

Table 6.8 Literature on the development of POE (Bordass & Leaman 2015) reproduced by permission of Bill Bordass and Adrian Leaman

Authors	Date	Area of Interest	Building Types	Country of Origin	Topic/Details (where appropriate full reference in list of references)
T. Markus et al.	1972	Appraisal techniques	Mainly schools	Scotland, UK	Building Performance, Building Performance Research Unit, Applied Science Publishers, Barking, London
Boudon, P.	1972	Lived-in architecture	Housing	France	Lived-In Architecture, Lund Humphries, London
Zeisel, J.	1984	Methods	Most building Types	US	Inquiry by Design, Tools for Environment-Behaviour Research, Cambridge University Press, Cambridge
Preiser, W. et al.	1988	Methods	Mainly offices	US	Post Occupancy Evaluation, Van Nostrand Rheinhold, New York
Hedge, A., & Wilson, S.	1987	Health in buildings	Offices	UK	The Office Environment Survey, Building Use Studies, London
Vischer, J.	1989	Environmental quality	Mainly offices	US	Environmental Quality in Offices, Van Nostrand Rheinhold, New York
Baird, G. et al.	1996	Methods and cases	Various	NZ	Building Evaluation Techniques, McGraw-Hill, New York
The Probe Team: Bordass B., Leaman A., & Cohen R.	2002	20 building case studies	Various	UK	Walking the tightrope: the Probe team's response to BRI commentaries *Building Research and Information*, Special Issue, March 2002 **30**(1)

(http://www.usablebuildings.co.uk/, look for the PROBE link on the right-hand side).

POE methods are quite varied (Leaman 2003), and Riley (2009) confirms that there are more than 150 POE techniques available worldwide, with 50 available within the UK.

Page 15 of the AUDE 'Guide to Post Occupancy Evaluation', prepared by Estates and Facilities at the University of Westminster and available at http://www.aude.ac.uk/resources/goodpractice/AUDE_POE_guide/, has a table of existing POE methods available (Table 6.9) (AUDE 2015).

More recently, broader approaches now apply POE to prehandover processes and in some instances link POE to early involvement of a range of parties, enabling some issues to be addressed before they become more difficult to manage. A number of authors have researched the connections between POE and the roles played in designing (Cook 2007), constructing (Williams et al. 2013), and managing buildings. Various parts of the construction process and their links to POE are highlighted in Williams et al. (2013), which focuses on 'Principal contractor involvement in post-occupancy evaluation in the UK construction industry'. Within this work, the researchers highlight the lack of contractor involvement in POE and the related issue of insufficient knowledge on how or when to conduct POE on a project.

To date, the group has delivered the following:

- Index for Policy Document and Guidance. Providing an overview of the policy, its structure and contents
- Updates for industry through various presentations
- Process maps for GSL and associated guidance documents, showing the link between the stages of BIM, its data drops, and those required for GSL
- Trial projects for the use of GSL that will also support the longer-term development of the means to implement the policy and ensure that it becomes embedded within government practice

In a recent move, the government has stated that GSL will apply to all central government projects by 2016 (BSRIA 2014).

6.6   Commissioning and technical assessment

A building's performance can be predicted, and this is often necessary when aspects such as the size of the building services required for heating, cooling, water usage, and ventilation are being considered. However, to be able to understand how reasonable these predictions are, some form of assessment of the true building performance is expected. This section looks at a number of the more common systems and techniques that can measure or assess that performance.

Commissioning and undertaking a technical assessment of a building and its design are linked. Commissioning can be seen as a process that will, as far as possible, enable the building to perform as the design intended, reduce energy usage, optimise occupant comfort, fault-find, and fix. A technical assessment tends to be less hands-on and would normally make recommendations rather than actually take physical steps to alter a building's performance. Technical assessments can take place when a building is still a design or be undertaken once a building is completed. The importance of both of these processes lies in their ability to step back from the usual construction professional's role of making sure the client, often the future building owner, is satisfied with the cost, the quality, and the handover date. In this way the sustainability of the building is championed through dispassionate eyes and the longer-term performance of the building is a goal of good commissioning and holistic technical assessments.

Commissioning of fixed buildings services is a stipulation of the UK's buildings regulations (Part L2A, 2013), Conservation of Fuel and Power of New Buildings Other than Dwellings, which came into force on 6 April 2014 (HM Government 2013b). It is advised that a commissioning plan be set out, and that plan is followed by either the contractors that supplied and fixed those services or by appropriate subcontractors. Additional support is provided by the CIBSE *Commissioning Code M: Commissioning Management* and other CIBSE documents, *TM31: Building Log Book Toolkit*, *TM39: Building Energy Metering*, and *TM46: Energy Benchmarks* (CIBSE 2003; 2006; 2008; 2009; 2014). Taking the guidance contained in these documents and the best practice requirement for a commissioning plan, it is logical to have a single person or team (depending on the size of the task) dedicated to managing the commissioning process. One of the major reasons for this is that since many of the separate parts of the building that require commissioning are often supplied and installed by different companies, the end result of the commissioning process could be a series of separately commissioned elements. This could further lead to a lack of continuity between each element and the building as a whole failing to function correctly. This is particularly difficult if the building's control systems are fully integrated and if an element of smartness (intelligence) within the systems has been installed. The balance between comfort and productivity alongside lowering energy use can also be part and parcel of the process of commissioning. The act of commissioning should also transfer up-to-date guidance to the building's log book for later use and, through this, encourage the building managers to keep the log book updated with further enhancements of hardware and software. Where some parts of the building services were originally specified but since the design phase were not implemented, the commissioning report could comment and suggest some specific upgrades when the budget allowed. Technical assessments can be varied in their purpose, scope, and use of technology. This can vary from a relatively simple analysis of existing energy billing information through to complex thermal assessments and simulations. One of the most effective can also be linked to an extension of the building energy and water use measurement and recording devices available for buildings.

6.6.1   Monitoring and metering

Monitoring the performance of our buildings can be associated with many of the methods and technologies introduced in this chapter. It is important that meaningful data is collected to ensure the longer-term performance of a building for the occupants, owner, maintenance suppliers, and the wider community. Monitoring is normally associated with the measurement and collection of numerical data but might also be associated with qualitative comments. The purpose of both numerical and qualitative data is to provide a true reflection of what is happening in the building over appropriate time periods. Some of the more qualitative elements of assessing building performance are covered within the 'Post-occupancy Evaluation' section of this chapter and within Chapter 7. The numerical elements, particularly those associated with the energy-related performance of a building, normally rely on devices that measure and sometimes record the energy performance over time: meters.

6.6.1.1   Why use meters?

Metering systems might, at first, not be the most exciting of technologies that can be associated with sustainable construction. However, the role that targeted data acquisition, recording, and display can play in proving and improving the performance of buildings should not be underestimated. Reasons for using meters are various, and the motivation for using them depends on whether an interested party is an occupier, owner, designer, maintenance contractor, or portfolio manager.

Occupier: Effective metering can enable an occupier to tune their behaviour to allow them to live the way they wish to but with the knowledge of how they can use different utilities to avoid energy and water waste and save money.

Owner: (particularly an owner who is providing a tenant with energy and water as part of a tenancy agreement) can use metering to pinpoint areas of potential overuse, or suggest ways in which the building can be reconfigured to reduce usage.

Designer: (and some owners) will need to have installed certain stipulated metering systems to comply with building regulations. If the building is to be put through a BREEAM assessment, the final rating will be influenced by the metering strategy.

Maintenance contractor: Data from metering systems, if timely and appropriately communicated, can allow contractors to detect faults or help predict a future problem. (There is a fine line between predictive maintenance and using metering data, but the overlap is great enough to allow maintenance contractors to be included in the broad range of interested parties.)

Portfolio manager: Enables an organisation that is responsible for a range of buildings to track, compare, and optimise the way in which people use those buildings and their technologies, both innovative and traditional in scope.

Building performance: It is advantageous to all that buildings perform as well as they can. By 'all' this infers that any reduction in the use of resources will be positive, whether this is a passer-by, a multinational company, or a government department.

Knowing how a building is performing is a prerequisite to comparing that performance against the predefined benchmarks of the performance of similar buildings, the performance predictions undertaken at early or late design stages, and the ability of separate aspects of underperformance to be examined in detail. For this to be known, a series of meters need to be used to measure and collect data in a form that can be easily read and acted upon. These meters come in many different guises and vary considerably depending on the task these meters perform, the complexity of the system being monitored, and the type of energy or resource being used. This is further complicated by the distinct differences between normal domestic-type meters and the inherent complexities of heavily serviced non-domestic buildings. It is therefore helpful to diagrammatically represent a metering, data capture, and analysis system that would broadly function for either a domestic or non-domestic building, as shown in Figure 6.4.

```
┌─────────────────────────────────┐
│          The Building           │
└─────────────────────────────────┘

┌─────────────────────────────────┐
│   Client/analysis team (the     │
│  individual, group or company   │
│  that requires or analyses the  │
│            data).               │
└─────────────────────────────────┘
```

```
┌─────────────────────────────────┐
│ Software interface that provides various │
│ functions for deep analysis/communications │
│ of trends, and comparisons based on one or │
│        multiple buildings        │
└─────────────────────────────────┘
```

```
┌─────────────────────────────────┐
│ Data collection/logging/storage/communication │
│   hub. This device takes the data, raw or │
│ otherwise from each of the separate meters and │
│ sorts and stores the data until software interface │
│       requests a download.       │
└─────────────────────────────────┘
```

```
┌─────────────────────────────────┐
│  Meters that are often located close to the │
│    energy type/consumed resource that │
│  measure at predefined intervals (sometimes │
│ continuously) the consumption of the energy │
│  type/resource and can communicate with the │
│       data collection facility.  │
└─────────────────────────────────┘
```

Figure 6.4 A representation of a building-related metering, data capture, and analysis system.

6.6.1.2 Simple domestic metering

Many domestic properties that have their metering system supplied and used for the benefit of the utility companies will not have any facility more sophisticated than a readout located at the point where the electricity, gas, or water enters the home. Whilst this can be read by the occupier/owner/ utility provider, this situation does not allow readings to be recorded or remotely communicated. One could take the opinion that leaving this situation as the status quo is acceptable as the meter's function adequately for the utility companies; they are reliable and are relatively fool proof as far as measuring quantities of usage, for which the company can then be paid. However, there are a number of built-in problems that make them less able to fulfil the wider functions that society will need if our buildings are to move to being sustainable. Those issues include remote reading and the amount of time and travel expended to collect meter readings for at least two occasions in each calendar year (the norm for most utility companies in many countries), which is large. Meter readings can be supplied by owners and/or occupiers, but this relies on accuracy and timings that are not within the control of the utility companies. The timing and extent of any remote readings can be decided by the utility company, thus ensuring more accurate billing and the possibility of pinpointing faults or misuse. The issue of external organisations, particularly ones with commercial interests, is a possible concern, and transparency and safeguards are two elements that need to be appropriately tackled before any regimes and systems of collection and sharing of data are finalised.

6.6.1.3   Storage of data

Just as many different storage techniques have been developed alongside the advances in computing systems, the methods of storing and metering data have also become diverse. Where data can be stored varies from inside the case of the meter, to removable cards, to remote storage via wireless or net connections. One of the major issues connected to the storage of data is the capacity of any storage system. Meters can be 'asked' to stop recording data when the memory is full, overwrite the data, or in some instances, where the system is comprehensive, automatically download the data to a remote destination, clear the 'on board' memory, and start recording data afresh.

6.6.1.4   Communication and analysis of data

For the meter to influence building operation, the data has to get to those people who can propose changes in light of any findings. Very rarely can anyone view a table of numbers and make wide-ranging decisions based on trends or patterns. This points to the need to present data in graphical format and to carefully select the variables that are going to be displayed. Obvious connections are often a good place to start: outside air temperature and the energy consumption of heating and cooling systems, and time-related relationships such as energy consumption in commercial buildings 'out-of-hours'. These can often pinpoint pieces of equipment that are either malfunctioning or operating at times when they would not normally be expected to operate. How to do this? The data can be assessed after the meter has been recording, through the use of historical data, or it can use real-time data to give an immediate snapshot of current consumption. The approach is normally different for these two types of data. Historical data is normally analysed using spreadsheet-type tools where longer-term patterns can be compared. Real-time data is often displayed in the form of a dashboard screen, adjustable to show the data a user needs to see. More simple devices have a built-in display allowing the user to scroll down through a series of different measured parameters.

6.6.1.5   Smart and intelligent metering systems

The term *smart meter* has been used in a range of situations and can refer to the process of measuring the usage of a commodity by extending this capability from the more basic recording usage functions to advanced systems that will perform calculations, show the user options, and anticipate the result of future demand. True smart metering also includes the ability to add some form of control of buildings services using the results of the metered values. This aspect of control is addressed in the section of this chapter titled 'Smart Metering and Control'.

The domestic smart meter can have very different capabilities to a smart meter that is designed to operate in a non-domestic or commercial building. Intelligence implies the meter would be expected to interact with the building, automatically controlling devices that may not even directly consume large amounts of energy but can reduce the overall energy usage. Here are some examples of these different characters within the greater fold of smart meters.

6.6.1.5.1   Domestic smart metering

Domestic smart meters are often typified by their use of a graphical display, wirelessly linked to a sensor connected to a clamp placed around the mains cable (Figure 6.5) or consumer unit within a domestic or domestic-scale building.

The displays vary in design, but most will give the occupier the ability to see the real-time energy usage, both in kWhs and in a monetary value of current and/or historical energy usage. Some systems use IAMs or 'individual appliance monitors' that add the ability to measure the energy from one appliance to another. This can be particularly useful in relation to behaviour change. (Please see Chapter 7 for more detail concerning the links between behaviour and sustainable construction.) A further development is to enable

Figure 6.5 Clamps around electrical cables.

utility companies to monitor usage remotely and inform their planning to ensure the stability of the UK grids and networks of energy and water. The next logical step is that utility companies develop the ability to turn off and on some of our less time-dependent energy usages, such as stored hot water production. In this way when the use of electricity is low, perhaps in the early hours of the morning, electrically operated water heaters can be operated between defined times, exactly when the utility company has spare generation capacity. This could be seen as an extension of the low tariff times that already exist. The key to this innovation will be the use of smart meters. This technology, however, raises a difficult issue of privacy and lack of end-user control, which has yet to be debated and resolved.

6.6.1.5.2   Non-domestic metering

While the metering of non-domestic buildings shares some basic similarities with domestic metering, the scale and variety of measurements combined with the complexities of technologies, control systems, and end users forces metering to be incorporated into systems. Building management systems (BMS), sometimes historically known as building energy management systems, are the current industry standard method of controlling and collecting data related to building services. There are also stand-alone metering systems that can be used alongside BMS. The Carbon Trust survey of metering in low-carbon buildings, 'Green Gauges', undertook an analysis of the lessons learned from 28 case studies, including education, retail, and mixed use residential buildings (Carbon Trust 2011). Whilst it appeared to be a good idea to use the data that was already available from a specified BMS, it found that in reality it was better to have a separate metering system that could be included as the relevant building service technologies were installed. The BMS, although very helpful for day-to-day usage, were found to be less capable of easy analysis of large amounts of historical data and not fully configured to produce easily accessed monitoring reports. This situation is likely to change, as first, facilities management sections of larger companies demand more easily obtained information from BMS and second, the technology related to the interfaces becomes more user friendly. More detailed information about the use of advanced metering and the related possible savings, both in cost and carbon, can be viewed in the Carbon Trust's publication *Advanced Metering for SMEs: Carbon and Cost Savings* (Carbon Trust 2007). The technical best practice can be found in CIBSE's publication *TM39: Building Energy Metering* (CIBSE 2009).

6.6.1.5.3   Commissioning of metering systems and its role in ensuring good data that is used effectively

Commissioning, or a correct handover of metering systems, is as important as the traditional commissioning of building services. If carefully designed, procured, and installed metering systems are not introduced to occupants/managers/FM/maintenance personnel, they are often overlooked and the many benefits greatly reduced.

The most effective methods of ensuring that commissioning and calibration take place appropriately are as follows:

1. When and how the commissioning will take place is built in to the metering scheme from a very early stage.
2. When some form of evidence that the commissioning took place is required, often an element of the building's handbook can have reference to this or perhaps a separate commissioning report for more extensive metering schemes/larger buildings.

The practicalities of any commissioning will vary according to the metering system being used, but some basic checks are likely to include the following:

- Check that the specified equipment agrees with the metering design.
- Calibrate any additional meters with any of the meters being used by utility companies, thus ensuring accuracy and that submetering totals agree with total consumption figures.
- Check that data flows also agree with conventional readouts.
- Ensure that the analysis of any data on screen agrees with printouts.
- Ensure that there is a schedule of maintenance checks that replicate some of the more appropriate commissioning functions to ensure long-term accuracy.
- Confirm training events that link in-house with off-site expertise
- Confirm logbook systems to pick up the above issues

6.6.1.5.4   Smart metering and control

Smart metering, as discussed, describes a number of different metering systems when applied to the measurement of utility use in buildings. The phrase 'smart metering' can be applied in a great many ways and, at the lowest level, might mean that the data from a building fire sensor is sent automatically back to a central location. For the higher classifications of smart metering, there will be two-way communication between sensors/meters and a certain degree of control as well as the analysis that can be undertaken on the data that comes from utility meters. At the highest level of smart metering, cooling humidification and processes beyond pure heating and ventilation can also be recorded, assessed, and controlled remotely.

One might ask, why do we need to have smart metering systems? With the ability to measure comes the ability to understand what is going on inside the building and to take appropriate action so that the building or the processes that take place in the building are monitored and controlled properly. Once this form of information is available, companies can then view their data and, where appropriate, set targets for future trends or compare existing performance against historical performance. Therefore, waste and wasteful practices can be identified. A number of Carbon Trust field trials carried out between 2004 and 2006 indicated that savings of up to 12% for energy spending would typically be available when smart meters

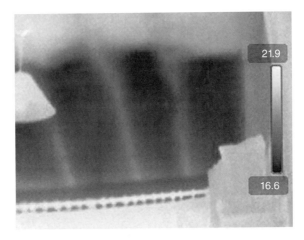

Figure 6.8 IR image of missing insulation (Photo by Steve Goodhew).

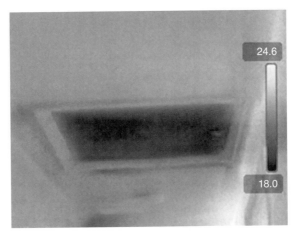

Figure 6.9 IR image of loft hatch (Photo by Steve Goodhew).

Figure 6.10 IR image of heat loss at roof ridge and chimney (Photo by Steve Goodhew).

235

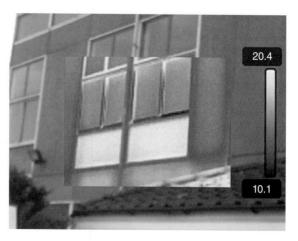

Figure 6.11 IR image of poor insulation levels in panel construction (Photo by Steve Goodhew).

Figure 6.12 IR image of heat loss through dormer window structure (Photo by Steve Goodhew).

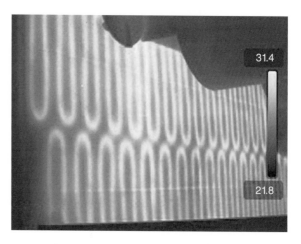

Figure 6.13 IR image of underfloor heating (Photo by Steve Goodhew).

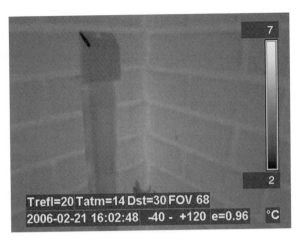

Figure 6.14 IR image of heat flow through brickwork joints (viewed from the outside) (Photo by Steve Goodhew).

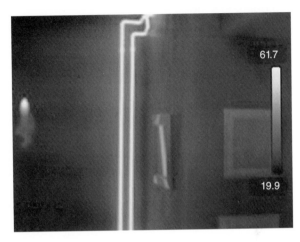

Figure 6.15 IR image of building services pipework (Photo by Steve Goodhew).

Training for thermographers is available at Levels 1, 2, and 3, where Level 1 is related to competence to undertake surveys with guidance, and 2 and 3 provide the training to manage the survey process and obtain more intimate knowledge of the instrumentation and processes. Level 2 is recommended as the minimum standard for any thermographer to take valid surveys that can be accepted for inclusion in BREEAM or other building assessment systems.

6.6.4 Air leakage testing

Many energy-efficient building designs rely partly upon high levels of insulation in a building's fabric. However, this approach can be reduced in effectiveness if the amount of air leakage in a building leads to unnecessary heat loss. Air leakage is uncontrolled air flow through cracks and gaps between parts of the fabric of a building, as opposed to ventilation, which is controlled air flow normally purposely provided through passive and active ventilators. To maintain good air quality in a building, the appropriate approach is to ensure that uncontrolled air leakage is minimised and adequate controlled ventilation is used to supply the building's and occupant's needs. To comply with the current UK regulations, adherence to maximum permissible leakage rates for buildings can be found within Part L (Conservation of Fuel and Power) of the UK Building Regulations 2013 (alongside methodologies contained in the ATTMA's TSL1) (HM Government 2013a). To comply with these regulations and guidance, the sources of uncontrolled air leakage need to be identified and sealed if necessary. These air leakage paths that run through a building's fabric can be difficult to identify, and gaps between different parts of the building's structure, finishes, services, and appliances can often be hidden. Therefore, the only appropriate method to test whether the building fabric is airtight to a stipulated standard is to measure the leakiness of a building's fabric as a whole.

6.7   Building simulation

One of the methods that many building design teams use to assess their designs prior to construction is to simulate their performance using building simulation software. The more comprehensive software will allow an accurate model of the building to be put through a year's worth of weather to record the model's performance. Different options relating to material/technology choice, building form, and usage can be factored into different simulations. The results from these can be used to compare different scenarios and strategies.

The range of software is great both in capacity (the different capabilities of the software) and complexity (the computational model that underpins the calculations that the software undertakes). The major two subcategories of software are those that rely on steady-state and dynamic algorithms. Crudely, the difference between the two is that the dynamic systems take into account the thermal storage capacity of the building as well as the insulation, whilst the steady-state software tends to base thermal calculations on thermal conductivity of the insulation of the building elements rather than any capacity effects. Dynamic systems are more comprehensive by nature but are also more complex to use and normally will need operators who are thoroughly trained and experienced. CAD drawings can be subtly altered to allow a building's dimensions and even materials to be input straight into a dynamic thermal simulation piece of software. As the impact of BIM is felt, the production of complete 'one-stop-shop' pieces of software that combine performance simulations, design evolutions, measurement data, management of the construction process, payments, ordering of resources and general sharing of important information are likely to evolve (A selection of simulation software types can be found at: http://apps1.

eere.energy.gov/buildings/tools_directory/ (USDOE 2015). The next few subsections describe the most commonly used simulations systems and packages, including some regularity-based ones that do not so much simulate as assess compliance.

6.3 Want to know more? Air leakage testing

ATTMA TSL1 (ATTMA 2010a) gives guidance relating to assessing the leakiness of a domestic building's fabric, whilst TSL2 gives guidance for the assessment of non-dwellings (ATTMA 2010b). The requirements of ATTMA for the measurement of the air permeability of buildings are based on BS EN 13829:2001, 'Thermal Performance of Buildings—Determination of air permeability of buildings—Fan pressurisation method' (BSI 2001) with various enhancements recommended by ATTMA. BS EN 13829:2001 has now been replaced by BS EN ISO 9972:2015 – 'Thermal performance of buildings. Determination of air permeability of buildings. Fan pressurization method' (BSI 2015). TSL1 quantifies air leakage as air permeability. This is the leakage of air ($m^3h^{-1}m^{-2}$) in or out of the building per square metre of building envelope at a reference pressure difference of 50 Pascals (i.e., $m^3h^{-1}m^{-2}$ @ 50 Pa) between the inside and outside of the building. This assessment of building envelope air leakage establishes a difference in pressure across the envelope and measures the airflow required to achieve that pressure difference. This is normally achieved by utilising variable flow portable fans which are temporarily installed in a doorway, or other suitable external opening. All HVAC plant is switched off and temporarily sealed prior to the test, as are passive ventilation sources. All doors and windows on the exterior of the air test envelope are closed before the test commences. The pressure differential caused by the fan is increased by the temporary fan until a building pressure of 50–100 Pa is achieved. The total airflow, normally measured in volume of air/second, through the fan and the building pressure differential created between the inside and outside of the building, is normally about 50 Pa. By adjusting the fan speed in steps of about 10 Pa and noting the air flow (Q) and the pressure (Delta p) at each step, the relationship between the air flow and the pressure for that building can be established:

$$Q = C \, (Delta \, p)^n$$

where C and n are constants that relate to the specific building under test. Appendix A of ATTMA TSL1 provides the necessary equations that will allow a reader to establish C and N and establish the total air flow required to achieve the reference pressure differential of 50 Pa. This airflow is then divided by the total building envelope area to provide the leakage rate in $m^3h^{-1}m^{-2}$ @ 50Pa.

Table 6.11 Building Regulation Requirements for air permeability and air change rates Part L 2010 (England and Wales), Part F1 (Northern Ireland), or Section 6 of the Domestic Handbook (Scotland)

Type of Building	Air Permeability $m^3h^{-1}m^{-2}$@ 50 Pa		Air Change rate h^{-1} @50 Pa
	Best Practice	Normal	
Dwellings			
naturally ventilated	5.0	7.0	—
mechanically ventilated	1.0	5.0	—
PassivHaus standard	—	<1.0	0.6

6.7.1 Steady-state simulation

Steady-state software is far simpler both in scope and ease of operation than dynamic systems. This has the disadvantage of not being able to accurately predict the operation of larger more complex buildings but does lend its use to early-stage design tools when more broad, less accurate solutions are needed. Both steady-state and dynamic algorithms have the ability to receive the dimensions, materials, and other fundamental data concerning a building and its site to calculate the building's performance over an average year's weather data. This information is used to produce a virtual 'model' that can be placed in different geographical locations for accurate reflection of local conditions. Occupancy patterns can also be tuned to help predict relationships between different ways of using a building. The unique ability of this form of assessment tool is that it allows the building's design to be changed in very fundamental and major ways and can test the building's worth. This can range from predicted energy use and internal comfort conditions to outputs from connected software that can predict the costs involved in building the project.

Of the two types of simulations the dynamic software is better suited to larger buildings, both through the accuracy it provides and through detail of separate variables. From the latter, the software can compute results that can influence the design of larger scale heating and cooling systems. However, the simpler steady-state systems can be useful when trying to quickly assess a building's conformance to regulations or benchmarks and can be exemplified by the UK regulation software SAP.

6.7.1.1 Standard Assessment Procedure (SAP) and Building Research Establishment Domestic Energy Model (BREDEM)

The SAP is the main assessment method used in large parts of the UK to assess and compare the energy and environmental performance of dwellings

(UK Government 2014b). SAP's calculations assess the energy consumption of a dwelling and the quantity of carbon dioxide (CO_2) emitted when assuming that particular interior temperatures and occupancy patterns will be maintained inside the dwelling. SAP was based on the Building Research Establishment Domestic Energy Model (BREDEM), which was first developed in the 1980s (BRE 2012). BREDEM considers energy use for heating, hot water, cooking, fixed lighting, and the use of domestic appliances. BREDEM provides a simple, reasonably accurate model with which to assess the fuel costs associated with domestic buildings and identify the causes of fuel poverty and condensation.

6.7.1.2 Simplified Building Energy Model (SBEM)

SBEM (BRE 2013) is a steady-state software tool developed by the BRE (the UK's Building Research Establishment) that allows the estimation of a nondomestic building's energy consumption. SBEM is used in connection with some UK regulations and the European Energy Performance of Buildings Directive. SBEM is also intended to be used alongside one of the UK Government past initiatives for encouraging building occupiers to invest in energy saving measures, The Green Deal. The main focus of the tool within these regulations is to estimate the rate of CO_2 emissions for new buildings in compliance with the thermal regulations related to buildings (Part L, England and Wales and equivalent Regulations in Scotland, Northern Ireland, the Republic of Ireland, and Jersey) (DCLG 2013). SBEM is also used at the time of sale or agreement to rent of non-domestic buildings to produce an Energy Performance Certificate.

6.7.2 Dynamic thermal simulation

Dynamic thermal simulation software systems often started as a single piece of software concentrating on the simulation or assessment of thermal flows through a building, in its fabric, services, and internal and external heat gains/losses. Many originated as tools to allow engineers to size and design buildings' services, particularly HVAC (heating, ventilation, and air-conditioning) systems. Over time these dynamic simulations also included the ability to assess energy usage, the performance of building fabric and form, the impacts of use, fire-related options, computer fluid dynamics, and legislative assessments. This allows research, consultancy, and teaching communities to use the systems with a single model at the core of a number of different simulation techniques.

6.4 Want to know more? Dynamic thermal simulation packages

Developed at Strathclyde University in the 1970s, ESP-r was one of the first simulation engines. The software can be obtained free of charge under an open source license and is designed to run on a Unix operating system http://www.esru.strath.ac.uk/Programs/ESP-r.htm (ESRU 2015). The ESP-r system has been developed since its inception in 1974 to simulate the energy and environmental performance of buildings. Models that are built in ESP-r can be exported to other assessments tools such as EnergyPlus and Radiance (photo-realistic visual simulations).

EnergyPlus is also free of charge and when used as the version downloaded from the US Department of Energy website, http://apps1.eere.energy.gov/buildings/energyplus/ (USDOE 2014), it uses a database front end that requires a good knowledge of the implications and choices related to this form of data input. There is a 'sketch-up' (Sketch-up is a 3D drawing package) and several other graphical interfaces that can meet a range of user requirements. EnergyPlus is constantly being developed and has additional modules for the analysis of environmental construction details such as green roofs, renewables, and more whole life elements, such as embodied carbon.

IES VE is one of the market leading dynamic simulation tools. The interface is graphical and being constantly developed. The software works through a variety of modules that are focused on a variety of specialist areas such as building regulations, photo realistic lighting, natural and forced ventilation, existing buildings, CO_2 emissions, and so forth (http://www.iesve.com/) (IES 2015). IES VE is also linking to BIM platforms and is developing the ability to assess whole communities and cities.

TAS from EDSL (http://www.edsl.net/main/) (EDSL 2015) is a very competent dynamic simulation tool. It has a graphical front end interface with some effective methods of dealing with windows and shading. The tool uses some packages developed for manufacturer data-backed modelling of building services, such as passive stack ventilation units, fan coils, and so forth.

Bentley's Hevacomp's suite of building performance products (http://www.bentley.com/en-GB/Products/Building+Analysis+and+Design/Hevacomp.htm) (Bentley 2015) support ISO, IEE, CIBSE, and ASHRAE standards and have been developed for more than a quarter of a century. They include mechanical and energy modelling, electrical design, and energy assessment.

Retscreen packages are aimed at renewables analysis (http://www.retscreen.net/ang/home.php) (NRCAN 2015) and have a long history of providing effective spreadsheet-based assessments that are both accessible and often updated regularly.

There are a great many other specialist assessment tools available, such as Pilkington Spectrum (a specialist package that focuses on glazing performance, http://www.pilkington.com/products/bp/downloads/

tools/spectrum/default.htm) (Pilkington 2015). It would be wise to seek advice from any part of the relevant design and/or construction team alongside viewing some reviews of these products before embarking upon what can be a lengthy process of installing the software and training any necessary operators.

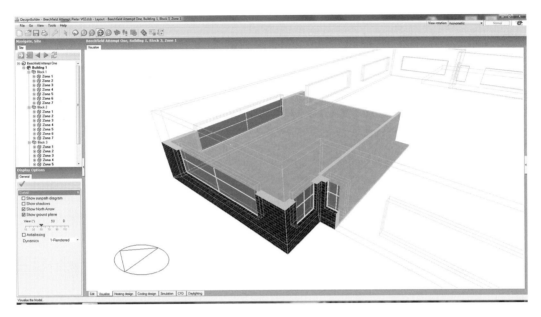

Figure 6.16 Example of a simulation model reproduced by permission of Prof. Pieter de Wilde.

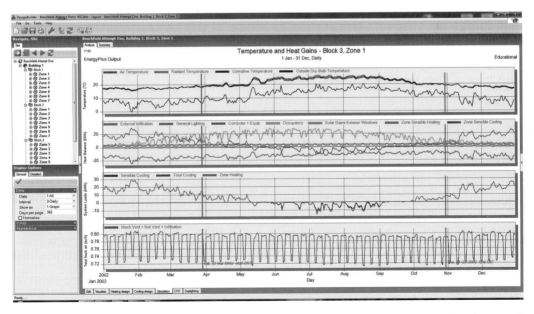

Figure 6.17 Example of results from a simulation model reproduced by permission of Prof. Pieter de Wilde.

6.7.3   Validation

The relation of the results from simulation software to the reality of the actual building is normally described as a form of validation for the software. This validation can be used to improve the reliability of the software but also to influence the accuracy needed for the data that is used to construct the software's virtual model. There are a number of published texts that advise and comment upon validation and also sensitivity analysis, including Lomas and Eppel (1992) and Lomas (1991).

Figure 6.16 show examples of a simple thermal simulation model of a domestic dwelling, and Figure 6.17 shows the results of a simulation for a model including various outputs such as temperature, ventilation, and so forth.

The above figure (Figure 6.17) shows the simulated response of a residential building with multiple zones over the period of a year. The simulation shows a series of temperature traces, including indoor air and radiant temperatures alongside external temperatures. The second graph down indicates the solar gain and other elements that contribute towards the heat balance of the property. The third graph down indicates the requirements for heating and cooling, and the final graph shows the air infiltration to the property.

6.8   Summary

Whilst assessment systems can aid the designer, site manager, and aftercare professionals in ensuring that buildings are sustainable, there are still the usual associated problems with any system or data stream. If building professionals do not use, interpret, communicate, and act in light of the results from the appropriate assessment system, then the work related to applying these assessments will be wasted. The possible perceived burdens of these systems will then not contribute to better buildings. Recalling the adage that 'buildings without people perform just fine', Chapter 7 addresses the cumulative actions of people and their behaviour and offers some insights into how human interactions impinge upon the success of any sustainable construction project.

References

ATTMA (Air Tightness Testing & Measurement Association) (2010a) *Technical standard TSL1. Measuring air permeability of building envelopes (dwellings)*, [Online], Available: http://www.attma.org/wp-content/uploads/2013/10/ATTMA-TSL1-Issue-1.pdf [20 Apr 2014].

ATTMA (Air Tightness Testing & Measurement Association) (2010b) *Technical Standard TSL2. Measuring air permeability of building envelopes (non-dwellings)*, [Online], Available: http://www.attma.org/wp-content/uploads/2013/10/ATTMA-TSL2-Issue-1.pdf [20 Apr 2014].

Aude (2015) *Guide to Post Occupancy Evaluation*, [Online], Available: http://www.aude.ac.uk/resources/goodpractice/AUDE_POE_guide/ [19 Aug 2015].

Baird, G., Isaacs, N., Kernohan, D., & McIndoe, G. (1996) *Building Evaluation Techniques,* McGraw-Hill, New York.

Bansal, P., & Bogner, W. (2002) *Deciding on ISO 14001: Economics, Institutions, and Context, Long Range Planning,* Elsevier, Kidlington.

BEAM (2015a) [Online], Available: http://www.beamsociety.org.hk/en_index.php [19 Aug 2015].

BEAM (2015b) BEAM Plus assessment tool, [Online], Available: http://www.beamsociety.org.hk/en_beam_assessment_project_1.php [27 Sept 2015].

Bordass, B., & Leaman, A. (2015) [Online], Available: http://www.usablebuildings.co.uk/Pages/Unprotected/CAT2015/UBTCAT1p212May15.pdf [19 Aug 2015].

Bordass, B., Leaman, A., & Cohen, R. (2002) *Walking the tightrope: The Probe team's response to BRI commentaries, Building Research and Information*, Special Issue, **30**(1).

Boudon, P. (1972) *Lived-in Architecture,* Lund Humphries, London.

BRE (2011) *BREEAM New Construction non-domestic buildings Technical Manual SD5073 – 2.0:2011,* [Online], Available: http://www.breeam.org/breeamGeneralPrint/breeam_non_dom_manual_3_0.pdf [31 Aug 2015].

BRE (2012) *Building Research Establishment Domestic Energy Model (BREDEM): A Technical Description of the BRE Domestic Energy Model Version 1.1,* [Online], Available: http://www.bre.co.uk/filelibrary/bredem/BREDEM-2012-specification.pdf [19 Aug 2015].

BRE (2013) *SBEM: Simplified Building Energy Model* [Online], Available: https://www.bre.co.uk/page.jsp?id=706 [19 Aug 2015].

BRE (2014) *BREEAM UK New Construction, Non-Domestic Buildings (United Kingdom)* Technical Manual SD5076: 4.0—2014, BRE London, [Online], Available: http://www.breeam.org/BREEAMUK2014SchemeDocument/ [9 Oct 2015].

BRE (2015) BREEAM case studies, [Online], Available: http://www.breeam.com [31 Aug 2015].

BSI (1992) BS 7750:1992 *Specification for Environmental Management Systems.* Superseded by BS EN ISO 14001:1996 Withdrawn.

BSI (2001) BS EN 13829:2001 *Thermal performance of buildings. Determination of air permeability of buildings. Fan pressurization method* Superseded, Withdrawn [Online], Available: http://shop.bsigroup.com/ProductDetail/?pid=000000000019983036 [25 Sep 2015].

BSI (2011) PAS 2050:2011 (PUBLICLY AVAILABLE SPECIFICATION) *Specification for the assessment of the life cycle greenhouse gas emissions of goods and services,* [Online], Available: http://shop.bsigroup.com/upload/shop/download/pas/pas2050.pdf [31 Aug 2015].

BSRIA (2012) BSRIA Topic Guides, Building Information Modelling, TG 4/2012.

BSRIA (2014) *BSRIA Soft Landings Framework* (BG 54/2014) [Online], Available: https://www.bsria.co.uk/services/design/soft-landings/free-guidance/ [25 Sep 2015].

Carbon Trust (2007) *Advanced Metering for SMEs: Carbon and cost savings, CTC713,* Carbon Trust, London. [Online], Available: http://www.carbontrust.co.uk/publications/pages/publicationdetail.aspx?id=CTC713 [2 Jun 2014].

Carbon Trust (2011) *Green Gauges, CTG037,* Carbon Trust, London. [Online], Available: http://www.carbontrust.co.uk/publications/pages/publicationdetail.aspx?id=ctg037 [5 Jun 2014].

CCI (2015) *KPIzone Centre for Construction Innovation (CCI)* [Online], Available: http://www.ccinw.com/kpizone/UserLogin/index.php?ResourceID [31 Aug 2015].

CEEQUAL (2015) [Online], Available: http://www.ceequal.com/assessment.html [19 Aug 2015].

CIBSE (2003) *CIBSE Code M: Commissioning Management*, [Online], Available: http://www.cibse.org/knowledge/cibse-cc/commissioning-code-m-commissioning-management.

CIBSE (2006) *TM31, Building Log Book Toolkit*, CIBSE, London. [Online], Available: http://www.cibse.org/knowledge/cibse-tm/tm31-building-log-book-toolkit [47 Oct 2015].

CIBSE (2008) *TM46 Energy Benchmarks*, CIBSE, London. [Online], Available: http://www.cibse.org/knowledge/cibse-tm/tm46-energy-benchmarks [47 Oct 2015].

CIBSE (2009) *TM39 Building Energy Metering*, CIBSE, London. [Online], Available: http://www.cibse.org/knowledge/cibse-tm/tm39-building-energy-metering [7 Oct 2015].

CIBSE (2014) *CIBSE Guide M: Maintenance Engineering and Management NEW 2014*, CIBSE, London. [Online], Available: http://www.cibse.org/knowledge/cibse-guide/cibse-guide-m-maintenance-engineering-management [7 Oct 2015].

CLG (2010) *Code for Sustainable Homes; Technical Guide November 2010*, [Online], Available: http://www.planningportal.gov.uk/uploads/code_for_sustainable_homes_techguide.pdf [7 Oct 2015].

Constructing Excellence (2009) [Online], Available: http://212.69.36.161/tools/sustainablehomes/ [31 Aug 2015].

Cook, M. (2007) *The Design Quality Manual: Improving Building Performance* Wiley-Blackwell, Oxford.

Cooper, B.A., Ahrentzen, S., & Hasselkus, B.R. (1991) *Post-occupancy evaluation: An environment-behaviour technique for assessing the built environment, Canadian Journal of Occupational Therapy* **58**(4), 181–188.

DCLG (2013) *Approved Document L - Conservation of fuel and power*, [Online], Available: http://www.planningportal.gov.uk/buildingregulations/approveddocuments/partl/approved [31 Aug 2015].

Ecohomes (2012) [Online], Available: http://www.breeam.org/page.jsp?id=21 [2 Jun 2014].

EDSL (2015) [Online], Available: http://www.edsl.net/main/ [31 Aug 2015].

ENSLIC (2015) *Energy saving through promotion of life cycle analysis in building* (ENSLIC BUILDING) [Online], Available: https://ec.europa.eu/energy/intelligent/projects/en/projects/enslic-building [31 Aug 2015].

EU (2015) *EU Eco-Management and Audit Scheme* (EMAS), [Online], Available: http://ec.europa.eu/environment/emas/index_en.htm [31 Aug 2015].

Hedge, A., & Wilson, S. (1987) *Health in buildings: The office environment survey, Building Use Studies*, London.

HM Government (2013a) *The Building Regulations 2013 Conservation of Fuel and Power L1A* [Online], Available: http://www.planningportal.gov.uk/uploads/br/BR_PDF_AD_L1A_2013.pdf [10 Apr 2014].

HM Government (2013b) *The Building Regulations 2013 Conservation of Fuel and Power L2A* [Online], Available: http://www.planningportal.gov.uk/uploads/br/BR_PDF_AD_L2A_2013.pdf [10 Apr 2014].

GRI (2014) Global Reporting Initiative, [Online], Available: https://www.globalreporting.org [10 Apr 2014].

IEA (2004) *International Energy Agency, Annex 31, A Survey of LCA Tools, Assessment Frameworks, Rating Systems, Technical Guidelines, Catalogues, Checklists and Certificates.* [Online], Available: http://www.iisbe.org/annex31/pdf/M_directory_tools.pdf [10 Apr 2014].

IES (2015) [Online], Available: http://www.iesve.com/ [31 Aug 2015].

ISO (2006) ISO 14040:2006 *Environmental management— Life cycle assessment— Principles and framework* [Online], Available: http://www.iso.org/iso/catalogue_detail?csnumber=37456 [31 Aug 2015].

ISO (2011) ISO 19011:2011 *Guidelines for auditing management systems* [Online], Available: http://www.iso.org/iso/catalogue_detail?csnumber=50675 [31 Aug 2015].

ISO (2013) ISO/TS 14067:2013 *Greenhouse gases—Carbon footprint of products—Requirements and guidelines for quantification and communication*, [Online], Available: http://www.iso.org/iso/catalogue_detail?csnumber=59521 [31 Aug 2015].

ISO (2015a) ISO 14001:2015 *Environmental management systems—Requirements with guidance for use*, [Online], Available: http://www.iso.org/iso/home/standards/management-standards/iso14000.htm [31 Aug 2015].

ISO (2015b) ISO 9001:2015 *Quality management systems—Requirements*, [Online], Available: http://www.iso.org/iso/catalogue_detail?csnumber=62085 [31 Aug 2015].

ISO (2015c) ISO 14000: *Environmental management series*, [Online], Available: http://www.iso.org/iso/home/standards/management-standards/iso14000.htm [31 Aug 2015].

ISO (2015d) [Online] available http://www.iso.org/iso/home/standards.htm [31 Aug 2015].

Leaman, A. (2003) *Post occupancy evaluation*, Paper presented at Gaia Research Sustainable Construction Continuing Professional Development Seminars.

Lloyds Register (2014) *ISO 14001 Revision*, Lloyds Register, Coventry [Online], Available: http://www.lrqa.co.uk/standards-and-schemes/iso-14001/ISO-14001-Revision.aspx#sthash.LGKCEGtZ.dpuf [2 Jun 2014].

Lomas, K.J (1991) Dynamic thermal simulation models of buildings: New method for empirical validation. *Building Services Engineering Research and Technology* **12**(1), 25–37.

Lomas, K.J., & Eppel, H. (1992) Sensitivity analysis techniques for building thermal simulation programs. *Energy and Buildings* **19**(1), 21–44.

Mark, L. (2014) It's official: Government to scrap Code for Sustainable Homes. *Architect's Journal* 20 March 2014. [20 Apr 2014].

Markus, T., Whyman, P., Morgan, J., Whitton, D., Maver, T., Canter, D., & Fleming, J. (1972) *Building Performance*, Applied Science Publishers, London.

Morledge, R., & Smith, A. (2013) Building Procurement, Wiley-Blackwell, Oxford.

NRCAN (2015) *RETScreen International, Natural Resources* [Online], Available: http://www.retscreen.net/ang/home.php [19 Aug 2015].

Pilkington (2015) [Online], Available: http://www.pilkington.com/products/bp/downloads/tools/spectrum/default.htm [19 Aug 2015].

Preiser, W., Rabinowitz, H.Z., & White, E. T. (1988) *Post Occupancy Evaluation*, Van Nostrand Rheinhold, New York.

RIBA (2013) [Online], Available: http://www.ribaplanofwork.com/ [19 Aug 2015].

RICS (2015) [Online], Available: http://www.rics.org/uk/knowledge/ska-rating-/ [31 Aug 2015].

Riley, M., Moody, C., & Pitt, M. A. (2009) *Review of the Evolution of Post-Occupancy Evaluation as a Viable Performance Measurement Tool*. BEAN Conference 2009, BEST Research Centre (Built Environment & Sustainable Technologies), Liverpool John Moores University, UK.

TNS (2014a) [Online], Available: http://www.naturalstep.org/ [2 Jun 2014].

TNS (2014b) [Online], Available: http://www.naturalstep.org/en/abcd-process [2 Jun 2014].

UK Government (2010) Code for sustainable homes: Technical Guidance—November 2010 [Online], Available: https://www.gov.uk/government/publications/code-for-sustainable-homes-technical-guidance [31 Aug 2015].

UK Government (2014a) [Online], Available: http://www.civilservice.gov.uk/networks/pam/property-asset-management-in-government/facilities-management/government-soft-landings-gsl [10 Apr 2014].

UK Government (2014b) *Climate change and energy—Guidance, Standard Assessment Procedure* [Online], Available: https://www.gov.uk/guidance/standard-assessment-procedure [19 Aug 2015].

UK Government (2015) [Online], Available: https://www.gov.uk/government/publications/2010-to-2015-government-policy-energy-efficiency-in-buildings/2010-to-2015-government-policy-energy-efficiency-in-buildings#appendix-7-code-for-sustainable-homes [31 Aug 2015].

UK Government BIM Task Group (2015) [Online], Available: http://www.bimtaskgroup.org/ [31 Aug 2015].

USDOE (2014) [Online], Available: http://apps1.eere.energy.gov/buildings/energyplus/ [10 Apr 2014].

USDOE (2015) [Online], Available: http://www.buildingenergysoftwaretools.com/ [19 Aug 2015].

USGBC (2015) *Leadership in Energy and Environmental Design* (LEED) V4 [Online], Available: http://www.usgbc.org/leed [31 Aug 2015].

Vischer, J. (1989) *Environmental Quality in Offices*, Van Nostrand Rheinhold, New York.

Williams, T., Bouchlaghem, D., Loveday, D., & Law, C. (2013) Principal contractor involvement in post-occupancy evaluation in the UK construction industry. *Facilities* **31**(1/2), 39–55.

Yetunde, A. (2014) *BREEAM UK New Construction 2014, Key updates from 2011*, Presentation to Ecobuild 2014, 6 March 2014. [Online], Available: http://www.breeam.org/filelibrary/Presentations/DeliveringSustainableBuildingsSlides.pdf.

Zeisel, J. (1984) *Inquiry by Design: Tools for Environment-Behaviour Research*, Cambridge University Press, Cambridge.

7. Behaviour, sustainable construction, and the performance gap

7.1 People, sustainable construction, and buildings

In the end, the buildings we construct need to suit the people who use, own, and maintain them. Whilst one can follow all the advice within the previous chapters, people design, commission, and use buildings and people are very diverse in their behaviour and needs. This chapter—whilst not suggesting that all the behavioural aspects of all the interactions between us and our buildings are easily condensed into 20 pages (that would be folly indeed)—offers a series of pointers that establish some of the evidence for the variances in building performance (the logical point of comparison between buildings) and suggests some of the theories and interventions that might be helpful to building professionals when considering whether human-related issues could be a major cause for a building 'not working'. Other issues, of more import to internal organisation, are also discussed, such as corporate social responsibility and the ethics of materials selection (also touched upon in the materials section in Chapter 4).

7.2 The building performance gap

A building that is procured, designed, built, and managed based on good practice and founded on the thinking extolled by the previous chapters would, one would think, perform in a sustainable manner. Unfortunately, best laid plans can go awry. The reality of what a building is and does, can, for many reasons, fail to match expectations, even if the processes that helped to produce it were well thought through. The major issues are, to what extent does the building performance fall short? in what areas? and will this lack of performance have long-, medium-, and short-term repercussions?

Sustainable Construction Processes: A Resource Text, First Edition. Steve Goodhew.
© 2016 John Wiley & Sons, Ltd. Published 2016 by John Wiley & Sons, Ltd.

To understand the linkages between sustainable construction and the (under-) performance of a building, it is now helpful to define what is meant by such a performance gap. This is a gap between the predictions made for and expectations of the building at the design stage and the actual, subsequent performance. Whilst the main focus is on a building's predicted use of resources, performance gaps can be measured in terms of the following:

- Comfort issues, overheating, cold spots, high levels of air movement
- Use of control systems
- Durability of the building fabric
- Acoustical issues

Other gaps might include the less tangible but nevertheless apparent issue of client and occupant satisfaction. This is echoed by an article in *Building* (magazine) by Neil May (2013):

> … usability and in what we term as 'delight', which covers issues such as daylighting, spacial layout and quality of finish. These are all important because buildings are more than just energy machines. They have a considerable impact on our well-being, culture and environment in many ways.

The Carbon Trust, however, views the performance gap to be, in essence, an energy-related mismatch. A participant from one of 28 case studies carried out by the Carbon Trust relating to the performance of buildings in use commented:

> For our case study projects, the operational energy use was up to five times higher than estimates during design. There is an opportunity to close this gap and save money and carbon.
>
> (Carbon Trust 2012c)

Turner and Frankel (2008) found that some of buildings that had been awarded LEED certificates performed at below the baseline.

This energy gap is a quantifiable amount; it can be measured, compared, and analysed, and measures can be put in place to retrieve the situation. Doing so enables the feed forward of lessons learned.

An event held by the International Building Performance Simulation Association (IPBSA) England at Plymouth University in 2014, entitled 'Bridging the Energy Performance Gap', alluded to performance beyond energy metrics. The symposium highlighted how crucial it was that performance gaps are identified, understood, and bridged: '… if the industry wants to deliver buildings that are robust, for instance in terms of "occupant proofing" or "climate change proofing", and which are engineered to adapt to changing use conditions' (IPBSA 2013).

However, the gap in performance is a gap between the projected energy performance of the building and the actual. This means that the way in which the projected performance is computed and defined is critical so that the projections do not cause the gap. There is a subtle difference between using estimates of performance from tools that are envisaged to give an accurate estimate of a building's performance and using estimates that come from tools that are meant for assessing compliance. Many tools that are used to measure compliance of buildings with legislation or guidance are not designed to give accurate, quantifiable predictions of future performance. Energy-related compliance tools often use steady-state algorithms to calculate their results, and these do not allow for many energy-related properties. However, the rationale for measuring compliance is very different from obtaining accurate predictions; compliance tools need to be simple, effective, and moreover give the same answer for the same input parameters. Therefore, predictions from compliance tools are unlikely to be either accurate or reasonable. This could be the source of some performance gaps.

7.3   Occupant behaviour: the performance gap

Whether performance gaps are due to the manner of predictions or not, there is no doubt that people are key factors in achieving the sustainability targets of a building. They are important in their involvement in the design, the choices they make at construction stage, and through the operational control level.

Occupants can significantly impact a building's actual energy use (Branco *et al.* 2004; Emery & Kippenham 2006; Pilkington *et al.* 2011), even in highly energy-efficient buildings. The variance in energy use across residential buildings is large: 4%–26% (Guerra-Santin *et al.* 2009; Van Raaij & Verhallen 1983). Pilkington *et al.* (2011) found that energy demand varied by a factor of 14 between six similar terraced passive solar homes. A building's energy use can be 40% above prediction due to the inefficient behaviour of the occupants who can leave windows open or heat rooms when not needed (Yu *et al.* 2011).

The behaviour of building occupants and the impacts they have on a building's performance can be simple or complex. Behaviour can affect decisions about the upgrading of a building's fabric. It can also influence the use of the building. What is perceived as 'best performance' differs for domestic and non-domestic buildings, and therefore the corresponding advice relating to maintaining performance and a sustainable approach are also different.

7.3.1   Domestic buildings

The environment, constraints, and opportunities related to achieving sustainability targets in domestic buildings differ in a number of fundamental ways to those of occupants of commercial buildings. Times and types of occupation are not limited to working hours: domestic buildings could be inhabited almost continuously by the same group of people, both day and night, such as by elderly inhabitants. Conversely, a busy working couple might have long periods of absence and only use their property as a place to sleep. Planning for very different occupation patterns for the same property is a challenge. The required internal environment can be very different depending on the type of occupant; sedentary occupants are more likely to require higher internal air temperatures than the busy working couple. This might lead to different strategies when specifying and using heating and control systems.

Occupants' autonomy to change their internal environments also varies. Commercial buildings that are located in city centre sites, sealed due to the pressures of external noise and the need for HVAC systems to perform effectively, are very different to domestic and other commercial settings in terms of their control of temperature and ventilation/dictatorial approach to occupants' temperature and ventilation. Domestic building occupants are normally free to alter their own building's controls and fabric openings to adjust the building's internal environment to suit their own requirements. This autonomy also stretches to the ability of owner-occupiers in domestic properties to upgrade their building's fabric and heating/cooling and control systems.

One of the major issues of predicting the performance of buildings, therefore, is the variance of the behaviour of different occupants and, furthermore, the lack of firm data that define these behaviours in a reliable fashion. Attempts have been made to model occupant behaviour with a view to defining occupancy patterns in energy simulation models (Richardson *et al.* 2008).

7.3.2   Occupant behaviour: rebound effects

Many energy efficiency improvements, particularly those undertaken in domestic properties, do not reduce energy consumption by the amount predicted. A phenomenon similar to the rebound effect was proposed as early as 1865 by the economist William Stanley Jevons when observing the improved efficiency of technologies using coal (Alcott 2005). The Jevons paradox proposes that more efficient technologies can increase, rather than decrease, the rate of consumption of any operational resources required. This has been taken to a present-day context and underpinned by the findings of a report produced by the Energy Research Centre's Technology and Policy Assessment based at the University of Sussex (Sorrell 2007). This report centres upon an observable phenomenon, the rebound effect, which describes some behavioural responses to improvements in the energy efficiency of a home. This effect can manifest itself in two ways, with direct and indirect effects. Direct rebound effects tend to stem from, for example, a householder increasing his/her demand for energy-related activities or services such as bathing, cooking, heating, or lighting after making some effort towards energy efficiency. This might be seen as a measurable effect when a property is insulated or a high-efficiency heating source is installed and the occupants decide to set their thermostat to several degrees above where it would be set in the average home or before the improvement. Possibly the occupants feel they no longer need to wear heavy jumpers indoors and wear the same clothing that they would in the summer months as they feel more comfortable that way. This is a direct consequence of the energy efficiency upgrade.

An indirect rebound, in contrast, is not related to the performance gap but does relate to wider sustainability issues. An indirect rebound effect may be observed where the householder saved energy and CO_2 by installing a renewable fuel source to the home but then undertook an action that transferred the monetary savings they accrued from the reduction in consumption of energy to fund a foreign holiday or purchase a piece of furniture. The extra energy that would be linked to the indirect gains, long haul flights and embodied energy to produce the furniture, produces greenhouse gas emissions that offset the gains in efficiency from the original upgrade. Some examples of direct and indirect rebound effects can be seen in Table 7.1.

A more detailed survey of estimating direct and indirect rebound effects for different UK socioeconomic groups was undertaken in 2013 (Chitnis *et al.* 2014) and included a number of building-related effects (Reproduced by permission of M Chitnis, SLRG):

- Insulating previously uninsulated cavity walls
- Topping up loft insulation to 270 mm
- Replacing existing boilers with condensing boilers
- Insulating hot water tanks to best practice (75 mm jacket)
- Replacing existing incandescent bulbs with compact fluorescent bulbs (CFLs)
- Replacing all existing lighting with LEDs
- Reducing average internal temperatures by one degree centigrade
- Direct heating

The same paper aligns the rebound effects for income alone against each of a range of commonly instigated measures. The results can be summed up in Table 7.2 (Chitnis *et al.* 2014).

Research based on data from the Netherlands examines the differences in occupant behaviour in homes and explores a direct rebound effect linked to the consumption of energy for space heating (Guerra-Santin 2013). This study

Table 7.1 Some examples of direct and indirect rebound effects

Phenomenon	Corresponding Indirect Rebound	Corresponding Direct Rebound
Factory saves considerable amounts of energy through energy efficiency Motorist purchases a new fuel-efficient vehicle Householder invests in a new LED lighting system, including lamps, luminaries, and controls.	The demand for energy will reduce, making it more likely that the price of energy will also reduce, leading to less of an incentive to reduce consumption. Drivers and householders may use the money they save to fly long distances for holidays, offsetting the energy efficiencies.	As the efficiencies of these projects/purchases/investments accrue, less energy will be needed to produce/travel/light, and the investor/consumer/householder will end up using more energy as it becomes more profitable to produce more goods or cheaper to travel farther or use higher internal lighting levels.

Table 7.2 Estimated rebound effects for an average household—income effects alone, ignoring capital costs

Measure	Rebound Effect (%)	Contribution of Direct Emissions to the Rebound Effect (% of total)
Cavity wall insulation	14.5	20.4
Loft insulation	14.5	20.4
Condensing boiler	15.2	20.4
Tank insulation	14.6	20.4
CFLs	15.3	29
LEDs	15.2	20.4
Efficient car	46.4	20.4
Temperature reduction	13.7	16.6
Car use reduction	28.1	8.5
Food waste reduction	77.4	24.1

found that occupants tended to accept higher indoor temperatures and to ventilate less after energy efficient improvements. The study also found that the improvement of the building's thermal properties and efficiency of heating systems still led to a reduction in energy consumption connected to heating even when the suspected rebound effect was taken into account.

A 2012 report focuses upon both the direct and indirect rebound effects in households calculated from seven measures that a typical UK household could take to reduce their consumption of electricity or heating fuels (Chitnis *et al.* 2012). Those seven ranged from introducing a new measure, such as insulation, where there was none before, to adding to an existing element and replacing a less efficient device.

- Cavity wall insulation in uninsulated cavities
- Topping up loft insulation to 270 mm
- Replacing existing boilers with condensing boilers
- Insulating hot water tanks to best practice (75 mm jacket)
- Replacing existing incandescent bulbs with compact fluorescents (CFLs)
- Replacing all existing lighting with LEDs
- Solar thermal heating

The report concluded that the rebound effects from the seven measures are in the range 5%–15%, depending upon the time period examined and assumptions used. It was estimated that direct rebound effects, such as increased energy consumption related to heating efficiencies, are rather smaller than indirect effects, owing largely to the small share of energy in total household expenditure. Calculations in other research support the findings. This work estimates the direct rebound effect by finding the difference between the following:

1. The calculated occupant and home-specific energy consumption of a home (calculated using degree day data for a reference year) and
2. The normalised measured consumption, and then dividing this by the calculated reference (1) (Hens *et al.* 2010).

7.1 Want to know more? The rebound effect

Greeninga, L., Greeneb, D., & Difiglioc, C. (2000) Energy efficiency and consumption, the rebound effect: A survey. *Energy Policy* **28**(6/7), 389–401.

Hens, H. (2010) Energy efficient retrofit of an end of the row house: Confronting predictions with long-term measurements. *Energy and Buildings* **42**(10), 1939–1947.

Sorrell, S., Dimitropoulos, J., & Sommerville, M. (2009) Empirical estimates of the direct rebound effect: A review. *Energy Policy* **37**(4), 1356–1371.

Herring, H., & Sorrell, S. (eds.) (2009) *Energy Efficiency and Sustainable Consumption: The Rebound Effect*, Palgrave Macmillan, 272 pp.

Review of Herring & Sorrell (2009): http://sspp.proquest.com/archives/vol6iss2/book.herring.html

This work took the normalised data from annual consumption data of 964 homes and constructed a direct rebound curve which can be (Hens *et al.* 2010), and was, used to predict heating consumption in the residential sector, taking into account the direct rebound effect. This work also suggests that energy prices can motivate occupants to initiate a direct rebound effect, as the work looks at two quite different fuels and prices.

7.3.3 Occupant behaviour: intervention to reduce sustainability

Occupants can intervene to affect the sustainability of the building, for example, with loft insulation. Often it is assumed that once insulation has been installed into a property, particularly a domestic property, that the benefits of lower rates of heat loss will continue uninterrupted over many years. This may be true for parts of a building where the insulation is relatively inaccessible and self-supporting. However, when a building occupant has access to parts of the building that have considerable amounts of insulation, occupier actions can influence the thermal performance of that building element. Fibre-based or looser-fill insulation can, by its nature, be susceptible to compression. This varies from product to product, but compression can either be permanent or temporary. The compression of these types of insulation will increase their density, which will, in turn, increase the material's thermal conductivity. Two pieces of recent research by the National Physical Laboratory (NPL 2014) and the Carbon Trust found that when compressed, loft insulation typically provides half the insulation levels advertised and that 82% of households use their loft for storage, in turn compressing any fibre or loose insulation (NPL 2013). At the time of writing the general recommended depth of loft insulation is 270 mm if the most commonly used fibre-based insulants are used. This will be greater in depth than the height of the majority of joists, which are often about

100 mm. Building occupants who use their lofts to place objects in or fit boarding over the joists will compress the insulation, so reducing its effectiveness. One solution is to provide a raised loft floor to provide loft storage, accounting for the extra depth required for current loft insulation depths whilst still supporting the loadings expected upon such a storage platform. There are proprietary solutions available that provide structural members that a platform can be fixed to. Other more traditional methods extend the ceiling joists at appropriate intervals. In theory, this reduces energy usage by maintaining the effective depth of the insulation. Whilst there is an obvious benefit for reducing heat loss through the ceiling of households, it does, however, reduce the temperature further in loft spaces. If stored items are susceptible to the cold and the increased moisture levels that can also follow, alternative storage would best be sought.

7.3.4   Occupant behaviour: understanding of sustainable technologies

An NHBC Foundation review focused on the behaviour of occupants in homes and explored the behavioural factors that impact energy use in the home (NHBC Foundation 2011). The review examines aspects such as guidance, feedback, and information to occupants, and the role of energy displays and smart meters. It also investigates the application of behavioural science theories for changing behaviour to energy use behaviours. The review found, from various evidence, that low-carbon buildings were underperforming and that their optimum performance was unlikely to be realised because occupants do not fully understand how to control the technologies effectively. The NHBC Foundation put this down in part to the inaccessibility of user interfaces and the misuse and/or misunderstanding of the systems installed. Their response focused upon the 'need to better understand occupant behaviour and how occupants interact with buildings (in particular, the user controls) as this can have a massive impact on the energy used and the comfort levels achieved' (NHBC Foundation 2011).

In 2006 the UK's BRE published findings from the UK government's Department of Trade and Industry (DTI) funded field trial of domestic photovoltaic installations (Munzinger *et al.* 2006). The report found that in many cases users are not sufficiently well informed to obtain the full advantage of their system. It was found that the main problems were as follows:

* Lack of understanding of the operation of the system
* Lack of documentation for reference and for transfer to new owners/ tenants
* Lack of a suitable display system that is informative whilst being easy to understand and accessible

One of the main conclusions of the report was the recommendation that forms of user engagement are included as an essential part of any project design.

But how can sustainable behaviours that will maximise the probability of achieving a successful sustainable project and building be encouraged?

7.4   Modelling using occupant behaviour

To allow simulation models to be used in situations where people are the major influence on model performance, the realism of assumptions relating to what people do and how they behave is very important. Occupant behaviour is a source of significant uncertainty in energy modelling. For example, predicted energy usage in a primary school was modelled as increasing by more than 150% from the lower to the higher values established by experts as representative of 'typical' occupant behaviour (Clevenger *et al.* 2013). The majority of researchers in this area of focus are trying to improve the accuracy of simulations and, in part, attempt to close the previously mentioned performance gap. An analysis of user behaviour and indoor climate in an office building in Kosovo was undertaken by Dibra and colleagues (2011). The study also correlates with most of the outcomes of the thermal performance (behavioural, comfort, satisfaction, and energy issues) of office buildings in other countries mentioned in the Dibra paper (Austria and Ghana). Generally it was found that some form of meaningful interaction between facility managers and the occupants of the building was needed with regard to the systems and their operation. Alongside this, the actual building occupancy and use patterns deviated significantly from typical assumptions in simulation studies. The UK's Zero Carbon Hub has previously published this ambition: 'From 2020, to be able to demonstrate that at least 90% of all new homes meet or perform better than the designed energy/carbon performance' (Zero Carbon Hub 2014). The implications of this target are profound and the behaviour of the occupants, appropriately modelled, can allow issues beyond different occupancy patterns to be investigated.

7.5   Behaviour in the design process

7.5.1   Behaviour and design

The behaviour of people both as individuals and as collective groups influences not only the direction and success of a construction project but also its performance and the longer-term functioning of the building. This influence can vary from the mindset a client has when a building is procured, to the influence of the designers and the actions of the contractors, to the buy-in of the aftercare personnel. Hoping that people will behave in the 'right way' but making no allowance or provision for this throughout a project may mean that the best performance (in terms of sustainable living) will not be realised.

Winston Churchill, a past UK prime minister, recognised the interrelationship between building design and how we act in and around our buildings when he said, 'We shape our buildings and afterwards our buildings shape us'

(Churchill 1943). (There is also a lesser-known quote from later in the same speech relating to the speed of construction. Churchill states, 'The last House of Commons, the one which was set up after the fire in 1834, was promised in six years and actually took 27 years'—interjection from Mr. Maxton [Glasgow, Bridgeton]—'We had not a bricklayer Prime Minister then' [Churchill 1943]).

Architects and town planners have been and are involved in influencing the behaviour of those who live and work in buildings. From Ebenezer Howard and the garden city movement (Letchworth and Welwyn) to the planners behind the waves of new towns in the UK (Basildon, Telford) and modernist architects such as Le Corbusier, many have been very aware of the way people live their lives and the impact of the built environment on them.

7.2 Want to know more? Town and city planning

This textbox will not present the detailed issues connected with wider town and city planning as these are beyond the scope of this work. However, here are some historical and additional texts that will guide readers if they wish to research this area of interest in more depth.

Howard, E. (1898) *To-Morrow: A Peaceful Path to Real Reform*, reproduced in various forms by a number of publishing houses, including a paperback by MIT, an e-book by Routledge, and a Kindle version in 2012.

Howard, E. (1902) *Garden Cities of To-Morrow*, S. Sonnenschein & Co., Ltd, London (Reprinted 1946, Faber and Faber, London).

UK Government (2015) Details of the UK New Towns Act 1946. Available on UKPGA website: http://www.legislation.gov.uk/ukpga/1946/68/contents/enacted.

Gutman, R. (2009) *People and Buildings*, Transaction, New Jersey.

Le Corbusier (1923) *Towards a New Architecture*, Dover Publications, New York, 1985.

As this text is being prepared, new communities and towns, many of which have an 'eco label' are being planned and built in a number of locations in the UK. One of these is 'Sherford', which is planned to be built in the English county of Devon, between the City of Plymouth and the upper part of South Devon. The development is to include 5,500 units (plus a further 1,500 post-2026), providing homes for a projected 15,000 residents alongside 84,000 m² of commercial and employment space. The planning of this sustainable, mixed-use development includes provisions for the following:

- affordable homes
- high-quality design
- community facilities
- walkable neighbourhoods
- high-quality public transport
- on-site renewable energy generation
- a community park (South Hams District Council, 2015).

Planning and architecture both offer an understanding of different types of space, zoning, culture, and people. To understand the connection between behaviour and planning on a larger scale and in a broader sense, some of the aspirations behind planning paradigms are needed. The previous textbox (7.2) notes some texts that informed the planners of new towns and the garden city movement in the UK and shows some details of a new community planned in the southwest of the UK.

Designing buildings and spaces for urban areas is intrinsically linked to people and their behaviour. Architecture and the analysis of building design in relation to people is a very well-documented specialist area, and this author does not presume to be able to adequately condense it into a chapter. However, there are a number of theorists and practitioners who advocate smaller changes that can be made to the design of buildings to improve the sustainability of a project by helping encourage certain behaviours from building occupants.

Avani Parikh has linked the relatively new behavioural paradigm of behavioural economics to new ways of looking at space and the environment (Parikh 2011). He emphasises that the built environment communicates to the occupants and they respond, sometimes consciously, sometimes unconsciously. His points tend to be made in relation to signage in buildings and the way the building can be designed to encourage what are perceived of as healthy activities, such as walking up the stairs rather than taking the lift. The positive aspects of behavioural traits are voiced as being most productive in persuading occupants to 'do the right thing'. This also has impacts that link the building's design with the design of the surroundings around the building, and the interaction between the two.

One building design that has employed some subtle but holistic concepts to promote its wider use and maximise the benefits of its construction was a health centre built in was Bromley by Bow, designed by Wyatt MacLaren Architects in 1998.

7.3 Want to know more? The Bromley by Bow Living Centre

The Bromley by Bow Healthy Living Centre was opened in 1998 as a partnership between a primary care team and a community development project. As with most health centres that intend to serve a community to enhance the health of, educate, train, and look after the welfare of the people who live near the centre, the building is about people and their interactions. Writing in the *British Journal of General Practice*, Sam Everington (2002) lists a number of elements that have been included in the design of the centre to increase the chances of people meeting. The entrance and waiting area are specifically designed to provide a welcoming environment. This area is free from multiple posters in the hope that health promotion is enacted simply through human communication. There are virtually no signs. This means that people ask for help, but this cultural difference intends this assistance to be offered more spontaneously.

The philosophy is that somebody wandering around looking lost is noticed, and help is then given. Whilst relatively common now, the waiting area has a food co-op, toy library, art group, baby massage, and a portrait artist. The building was designed for people to bump into each other as often as possible; the gallery kitchen, the community café, or the open reception area are all key to good communication. The reception is for every activity in the centre. The receptionist is more 'gate opener' than 'gate keeper'. There are no barriers or screens across the desk, which often only raise tensions. The centre is designed in a way that enables people to use as many spaces as possible. This creates real ownership, possibly helping to reduce vandalism to a very low level.

7.6   Behaviour and sourcing materials

7.6.1   Selection of materials: ethical standards

The selection of materials by a building's designer or contractor's specifier/ buyer is a part of the decision-making process that can influence the sustainability of a project. One particular service from the UK's BRE Global is GreenBookLive (BRE 2015): an online database used to identify products and services that can assist in the reduction of negative impacts on the environment. It brings together listings from a number of organisations (including a certain number from BRE) and offers information on 'green' products and services. Whilst the database is easy to use, the number of products in the database at time of writing was as not as large as expected, only six suppliers of insulation, for instance.

7.6.2   BES 6001 Issue 3

This standard sets out the requirements for a product to be classified as responsibly sourced. The requirements of BES 6001 (BRE 2014) provide a framework against which all construction products may be assessed. (The technical relationship between sustainable construction and this standard from a materials point of view is discussed in Chapter 4). The product can be assessed according to the following criteria: biodiversity, chain of custody, constituent material(s), environmental stewardship, ethical principles, responsible sourcing, senior management, chain and supply chain organisation(s), sustainable development, traceability, environmental declarations, lifecycle assessment (LCA)–based hierarchy, and water extraction. The standard also requires that any organisation adhering to the standard's aims should have a written policy approved by the organisation's senior management, to ensure compliance to the principles of responsible sourcing. The standard links the organisation's quality procedures to BS

9001 (ISO 2015), implying that BES 6001 would be carried out to a good standard and with confidence. Whilst BES 6001 is overarching and brings together the aspects of many other standards, there still needs to be a more comprehensive sign-up of companies. Below is some advice from BES 6001 relating to the use of renewable materials that illustrates how one particular section of the standard functions:

> The organisation shall establish a policy, supported by a documented management system, for the efficient use of constituent materials, to address the following issues, as appropriate to the product under assessment:
>
> - Use of renewable and/or abundant materials over non-renewable and/or scarce materials
> - Material resource efficiency—using less material to achieve a given function
> - Reuse of materials
> - Use of recycled materials and/or by-products (as defined in WRAP's Calculating and Declaring Recycled Content in Construction Products)
> - Use of recyclable materials
>
> (BRE 2014, reproduced by permission of BRE)

From the information that the company supplies, a rating can be made against each criteria and a final assessment can produce a rating of Excellent, Very Good, Good, or Pass. Pass would fulfil all the compulsory criteria and a nominal number of the more optional elements, whilst an Excellent would, logically, excel in all.

All the above are logical potential actions that fit within the theories and concepts of sustainable construction. What is probably most relevant is the way in which the standard might be policed and/or the way in which the public/commerce react to those who refuse to sign or disregard the standard's detailed requirements. Behaviourally, designers and specifiers can choose to aim for an Excellent or a Pass. What is the reputation of an institution if it is 'always known to aim for a pass', and what are the consequences of this, for example, in repeat business from discerning clients? This has implications for achieving sustainability targets and for the business of construction.

7.7   Sustainability and the business of construction

Construction companies, partnerships, and traders fulfil a number of roles, but one aspect that drives all their behaviours is the need to make a profit. This relates to sustainability and sustainable construction in many ways. The following section explores this aspect of profitability and its relationship with resource efficiency.

7.7.1   Resource efficiency

In a Department of Energy and Climate Change (DECC) business survey undertaken in late 2012 (EDIE 2012), 423 respondents ranked the importance of resource efficiency, energy, waste, procurement, carbon footprint, water, carbon emissions, transport, carbon emissions from buildings, and materials: overall carbon footprint. The DECC survey found that 83% of businesses believed that resource efficiency would become even more important to their business within the following two years. The result of this survey can be in part explained by the energy price increases listed by DECC in its *Energy Trends* for 2012 (UK prices). Between 2006 and 2011, in cash terms, gas prices increased by 17%, electricity prices by 26%, coal prices by 83%, heavy fuel by 120%, and crude oil prices by approximately 80% (HM Government 2012). This has driven many businesses to introduce targets to reduce energy and resource use and/or take waste reduction into account. The introduction of mandatory reporting of carbon emission levels from the largest companies and waste producers was introduced in April 2013 and is likely to have some longer-term influences. These include influencing the carbon usage and waste outputs in supply and value chains. In the longer term this will probably mean that the small companies that supply the larger ones will be pressurised to follow suit and report/reduce their carbon and waste. One could argue that all businesses should increase their operational efficiency; however, the likely impacts of a combination of regulations and difficult trading conditions have moved this efficiency from a position of 'we would like to if we can' to 'we must'.

7.7.2   Corporate social responsibility

Corporate social responsibility (CSR), one of the key elements of social sustainability, is now in place at a policy level in around 70% of the UK's extremely large and nearly 60% of the large or very large businesses. This is complemented by 67% of the extremely large companies having a designated board director who is responsible for the CSR elements of the company (Grant Thornton 2014). What has driven this behaviour? A number of less tangible yet still influential drivers influence this increased focus on CSR. For example, companies that have the ability to report to their employees, clients, and customers that their CSR policy is backed up by tangible actions can benefit in a number of ways. This should enable companies to recruit, satisfy, sell, and maintain business relationships, all of which are good for business. Lasting business relationships built on the knowledge that both sides of a partnership have committed to a more ethical way of doing and running their businesses tend to endure. Often businesses can guarantee their supply of materials and components through the trust gained from a more equitable trading environment.

Some legal elements relating to CSR are becoming increasingly important, and those businesses with foresight are looking at this issue as a driver for embedding CSR more effectively within their organisation. Some unsavoury publicity, possibly linked to the way employees are treated or questionable investment strategies, might also be traced back to a lack of focus on CSR, both on a corporate and individual level.

Many businesses are run with a combination of risk-tolerant and risk-adverse directions. Companies will normally try to take steps to quantify and analyse a risk before taking steps to reduce that risk or accept a plan to minimise the risk (often termed as part of a risk register). Sustainability can act as a prompt, encouraging efficiency within both commercial organisations and supply chains. This is likely to cover the overall risk for all connected parties within construction, where 86% of all employees work for small and medium enterprises (Jaunzens 2001).

The risk associated with the influence of climate change will play a part in the psyche of business behaviour over the coming years. Uncertainty related to extreme weather events cannot help but encourage businesses to look at climate adaption measures alongside climate mitigation. According to the report 'Why are business leaders prioritising sustainability?', 'by 2032 four times as many businesses could be at risk of flooding in the UK' (EDIE 2012). Drier summers alongside variations in water supply will also impact some construction businesses. Prolonged temperatures can cause difficulties for those getting to work and influence energy bills. Cold weather, often a normal reason for changes within the construction contractual landscape, may well also influence other issues beyond the daylong delays this normally causes. The ability for construction companies to be able to continue to safely satisfy clients despite the impacts of climate change legislation and customer drivers will largely depend on their ability to work and trade in a more sustainable manner. Buildings that are commissioned to be sustainable by companies that act with sustainable behaviours throughout the business process are more likely to provide value for their clients and therefore are more likely to be profitable.

7.8 Commissioning

The process of commissioning building services has traditionally been to ensure the safety and the internal environmental condition of the building to promote good comfort. Energy efficiency and emission reduction drives commissioning just as much as those traditional values. Chapter 6 describes and analyses the commissioning process along with metering and its relationship to sustainable construction. Commissioning does have a relationship with the perceived and actual requirements of occupants and is demanded by Part L of the UK building regulations in relation to non-domestic buildings.

The behaviour of owners and operators can be influenced by communicating the economic benefits of commissioning. Designers can be acquainted with the improved internal environmental conditions along with the probable improved enthusiasm of the buildings' occupants. The reputational benefit to equipment suppliers and specialist installers of effective commissioning (which ensures that all components and systems meet the criteria in any specifications and provide optimum energy and environmental performance) is a powerful driver (Noye et al. 2013).

> ### 7.4 Want to know more? Ceredigion Council and CIBSE Commissioning
>
> The Ceredigion Council building uses mixed mode measures, including a natural ventilation strategy combined with solar shading designed to reduce solar gain at particular times of the year. The natural ventilation uses motorised top-lights in each window in the office areas and motorised louvres in the atrium to promote stack ventilation. There was initially a problem with the atrium louvres driving fully when the window top-lights only opened a small amount. Feedback from occupants helped to identify and solve the problem.
>
> 'A well designed building user guide is something people can return to and new occupants can use it as a reference' (Carbon Trust 2012b).
>
> The UK CIBSE Commissioning Codes are invaluable in ensuring that a building's services are ready for long-term use, and they cover the following areas:
>
> * Code A: Air distribution systems (1996, confirmed 2006)
> * Code B: Boilers (2002)
> * Code C: Automatic controls (2001)
> * Code R: Refrigerating systems (2002)
> * Code W: Water distribution systems (2010)
> * Code L: Lighting (2003)
> * Code M: Maintenance engineering and management (2014)

A case study from the Carbon Trust document *Making Buildings Work: Lessons Learned from Commissioning Low Carbon Buildings* (Carbon Trust 2012a) looks at the role occupants can play in commissioning.

7.9   Facilities management

Constructing sustainable buildings is not a guarantee of achieving the performance levels that are normally associated with such buildings. As many sources state, the underachievement of buildings and particularly sustainable buildings to match their predicted performance can be, in part, reversed through good building and facilities management (FM). Indeed, an integrated commissioning period, handover, and planned FM strategy through a system such as the previously mentioned soft landings approach is a vital element of achieving expected levels of performance and satisfaction.

FM and maintenance decisions can influence the level of the performance gap. However, the aspects of organisational behaviour change can also be influenced by the actions of FM. Much of the burden of dealing with ongoing legislation relating to energy performance and the compliance of buildings (see Chapter 5 on energy compliance certificates) falls on FM professionals working with specific buildings or groups of buildings. Their advice and strategic plans can influence the behaviour of organisations, their actions, investments, and ways of working and use of their buildings. Azizia and others examined the ways in which FM professionals implement energy-efficient measures and concluded that many sustainable technologies could be inefficient during the operation of buildings. They also found that behavioural measures that targeted the engagement of building occupants in reducing energy usage vary in effectiveness (Azizia *et al.* 2014), emphasizing that targeting one set of occupants doesn't guarantee applicability in other instances.

7.9.1 Non-domestic or commercial buildings

As part of a DECC rapid evidence assessment, a group of UK-based researchers recently produced an energy-related conceptual framework related to influencing decision making in non-domestic buildings. The framework provides an outline theory of organisational behaviour and behaviour change and seeks to integrate insights from organisational theory, sociology, and economics. The assessment has a number of findings, and some of these (which have been decategorised from the quoted text) include:

7.9.1.1 Findings more generic in nature

- Differences between organisational energy behaviours are strongly linked to size and sector.
- Energy efficiency strategies differ across organisations and reflect their different motivations.

7.9.1.2 Findings more specific in nature

- Business investments in energy efficiency generally appear to require very high rates of return, in some circumstances much higher than other investments with comparable risks.
- The strategic value of energy efficiency (conferring competitive advantage) may be the key influence on whether investment in efficiency will take place rather than profitability.
- Non-energy benefits of energy efficiency, such as improved public image or comfort for staff, are critical to raising the strategic value of energy efficiency.
- As energy use is often invisible to management, energy management systems have a role to play in ensuring it becomes visible via monitoring, targeting, and reporting.
- The most successful strategies to deliver lasting change in workplace behaviours use a combination of technology change, feedback to users, and norm activation.
- Energy behaviours are highly diverse but also patterned and linked in systematic ways to the size of an organisation, its sector, subsector, and the local and national context (Banks & Redgrove 2012).

An interesting study that focused upon occupants in commercial buildings in Botswana and South Africa (Masoso & Grobler 2010) is 'The Dark Side of Occupants' Behaviour on Building Energy Use'. The research found that a larger proportion of the total energy consumed in each building was being used outside of working hours (56%) compared with during working hours (44%). The work established that this could largely be ascribed to occupants' use of lighting and equipment and related back to the internal environment of a building that requires cooling rather than heating.

7.10   Summary

At all points within the procurement, from the design, construction, commissioning, and use to the replacement of a building, the behaviour of people both as individuals and as collective groups influences not only the success and direction of a construction project's performance but also the longer-term functioning of the building. An understanding of these behaviours can help provide and maintain the effectiveness of sustainable buildings and the sustainability of the construction process. The next chapter takes a close look at a more detailed level of the selection and installation of sustainable innovations and technologies. Behaviour, as ever, plays its part in ensuring that the installation and handover of these technologies is successful for the client and the construction team.

References

Alcott, B. (2005) Surveys; Jevons' paradox. *Ecological Economics* **54**, 9–21. [Online], Available: http://www.blakealcott.org/pdf/sdarticle.pdf [4 Apr 2014].

Azizia, N.S.M., Wilkinsona, S., & Fassmana, E. (2014) Management practice to achieve energy-efficient performance of green buildings in New Zealand. *Architectural Engineering and Design Management* **10**(1/2), 25–39.

Banks, N., & Redgrove, Z. (2012) *What are the factors influencing energy behaviours and decision-making in the non-domestic sector? A Rapid Evidence Assessment*, Centre for Sustainable Energy (CSE) and the Environmental Change Institute, University of Oxford (ECI) for DECC, [Online], Available: https://www.gov.uk/government/uploads/system/uploads/attachment_data/file/65601/6925-what-are-the-factors-influencing-energy-behaviours.pdf [4 Apr 2013].

Branco, G., Lachal, B., Gallinelli, P., & Weber, W. (2004) Predicted versus observed heat consumption of a low energy multifamily complex in Switzerland based on long-term experimental data. *Energy and Buildings* **36**(6), 543–555.

BRE (2014) *BES 6001 Issue 3, Framework Standard for the Responsible Sourcing of Construction Products BRE Environmental & Sustainability Standard*, [Online], Available: http://www.greenbooklive.com/filelibrary/responsible_sourcing/BES-6001-Issue-3-Final.pdf [19 Aug 2015].

BRE (2015) Green book live, [Online], Available: http://www.greenbooklive.com/ [19 Aug 2015].

Carbon Trust (2012a) *Making buildings work: Lessons learned from commissioning low carbon buildings, CTG051*, Carbon Trust, London. [Online], Available: http://www.carbontrust.com/media/81377/ctg051-making-buildings-work-commissioning-low-carbon-buildings.pdf [10 Apr 2014].

Carbon Trust (2012b) *A natural choice, natural ventilation, CTG048*, Carbon Trust, London. [Online], Available: http://www.carbontrust.com/media/81365/ctg048-a-natural-choice-natural-ventilation.pdf [10 Apr 2014].

Carbon Trust (2012c) *Closing the gap, sharing our experience series, CTG047*, Carbon Trust, London. [Online], Available: http://www.carbontrust.com/media/81361/ctg047-closing-the-gap-low-carbon-building-design.pdf [10 April 2014].

Chitnis, M., Sorrell, S., Druckman, A., Firth, S., & Jackson, T. (2012) *Estimating direct and indirect rebound effects for UK households, SLRG Working Paper 02-12*, Sustainable Lifestyles Research Group, University of Surrey, Guildford. [Online], Available: http://www.sustainablelifestyles.ac.uk/sites/default/files/publicationsdocs/slrg_working_paper_01-12.pdf [19 Dec 2013].

Chitnis, M., Sorrell, S., Druckman, A., Firth, S., & Jackson, T. (2014) Who rebounds most? Estimating direct and indirect rebound effects for different UK socioeconomic groups, *Ecological Economics* **106**, 12–3.

Churchill, W. (1943) House of Commons rebuilding. *House of Commons Hansard Debates* **28**(393), 403–473.

Clevenger, C., Haymaker, J., & Jalili, M. (2013) Demonstrating the impact of the occupant on building performance, *Journal of Computing in Civil Engineering* **28**(1), 99–102.

Dibra, A., Mahdavi, A., & Koranteng, C. (2011) An analysis of user behaviour and indoor climate in an office building in Kosovo, *Advances in Applied Science Research* **2**(5), 48–63.

EDIE (2012) *Why are business leaders prioritising sustainability?* Environmental Data Interactive Exchange (EDIE), East Grinstead. [Online], Available: http://www.edie.net/downloads/Why-are-business-leaders-prioritising-sustainability/5 [3 Mar 2013].

Emery, A., & Kippenham, C. (2006) A long term study of residential home heating consumption and the effect of occupant behaviour on homes in the Pacific Northwest constructed according to improved thermal standards, *Energy*, **31**(5), 677–693.

Everington, S. (2002) Bromley by Bow, *British Journal of General Practice* **52**(476), 253.

Grant Thornton (2014) Corporate social responsibility: Beyond financials, *Grant Thornton International Business Report 2014*, [Online], Available: http://www.grantthornton.global/globalassets/1.-member-firms/global/global-assets/pdf/corporate-social-responsibility.pdf [9 Oct 2015].

Greeninga, L., Greeneb, D., & Difiglioc, C. (2000) Energy efficiency and consumption, the rebound effect: A survey, *Energy Policy* **28**(6/7), 389–401.

Guerra-Santin, O. (2013) Occupant behaviour in energy efficient dwellings: evidence of a rebound effect, *Journal of Housing and the Built Environment* **28**, 311–327.

Guerra Santin, O., Itard, L.C.M., & Visscher, H.J. (2009) The effect of occupancy and building characteristics on energy use for space and water heating in Dutch residential stock. *Energy and Buildings* **41**(11), 1223–1232.

Gutman, R. (2009) *People and buildings*, Transaction, New Jersey.

Hens, H. (2010) Energy efficient retrofit of an end of the row house: Confronting predictions with long-term measurements, *Energy and Buildings* **42**(10), 1939–1947.

Hens, H., Parijs, W., & Deurinck, M. (2010) Energy consumption for heating and rebound effects, *Energy and Buildings* **42**, 105–110.

Herring, H., & Sorrell, S. (eds.) (2009) *Energy Efficiency and Sustainable Consumption: The Rebound Effect*, Palgrave Macmillan, 272 pp.

HM Government (2012) *Energy Trends, A national statistics publication, June 2012, Special feature—Industrial energy prices* DECC, HM Gov, London. [Online], Available:

https://www.gov.uk/government/uploads/system/uploads/attachment_data/file/65929/5632-industrial-energy-prices-et-article.pdf [3 Mar 2013].

Howard, E. (1898) *To-Morrow: A Peaceful Path to Real Reform*, reproduced in various forms by a number of publishing houses, including a paperback by MIT, an e-book by Routledge, and a Kindle version in 2012.

Howard, E. (1902) *Garden Cities of To-Morrow*, S. Sonnenschein & Co., Ltd, London (Reprinted 1946, Faber and Faber, London).

IPBSA (2013) Bridging the Energy Performance Gap Symposium, Plymouth University, 25 October 2013, [Online], Available: http://www.ibpsa.org/Newsletter/IBPSANews-24-1.pdf (p. 33) [7 Oct 2015].

ISO (2015) ISO 9001:2015 *Quality management systems: Requirements*, [Online], Available: http://www.iso.org/iso/catalogue_detail?csnumber=62085 [31 Aug 2015].

Jaunzens, D. (2001) *Influencing small businesses in the construction sector through research*, BRE, London. [Online], Available: http://projects.bre.co.uk/sme/Download/Journ.PDF [4 Apr 2014].

Le Corbusier (1923) *Towards a New Architecture*, Dover Publications, New York, 1985.

Masoso, O.T., & Grobler, L.J. (2010) *The dark side of occupants' behaviour on building energy use, Energy and Buildings* **42**(2), 173–177, [Online], Available: http://www.sciencedirect.com/science/article/pii/S0378778809001893 [19 Dec 2013].

May, N. (2013) Bridging the performance gap, *Building*, 14 October 2013. [Online], Available: http://www.building.co.uk/bridging-the-performance-gap/5062092.article [19 Dec 2013].

Munzinger, M., Crick, F., Dayan, E.J., Pearsall, N., & Martin, C. (2006) *Domestic photovoltaic field trials: Good Practice Guide. Part II System Performance Issues*, HM Government, London.

NHBC Foundation (2011) *Foundation review: The impact of occupant behaviour and use of controls on domestic energy use* (NF38).

Noye, S., Fisk, D., & North, R. (2013) Smart systems commissioning for energy efficient buildings. CIBSE Technical Symposium Liverpool, 11–12 April 2013.

NPL (2013) Squashed loft insulation 50% less effective, *NPL News*, [Online], Available: http://www.npl.co.uk/news/squashed-loft-insulation-50-per-cent-less-effective [4 Apr 2014].

NPL (2014) Teddington, *NPL*, [Online], Available: http://www.npl.co.uk [23 Jun 2014].

Parikh, A. (2011) *Poster; Can Architects Nudge People into healthy action? Behaviour, Design and wellness*. [Online], Available: http://www.inudgeyou.com/wp-content/uploads/2012/09/Avani-Parikh-poster1.jpg [13 Dec 2013].

Pilkington, B., Roach R., & Perkins, J. (2011) Relative benefits of technology and occupant behaviour in moving towards a more energy efficient, sustainable housing paradigm, *Energy Policy* **39**(9), 4962–4970.

Richardson, I., Thomson, M., & Infield, D. (2008) A high-resolution domestic building occupancy model for energy demand simulations, *Energy and Buildings* **40**(8), 1560–1566.

Sorrell, S. (2007) *The Rebound Effect: An Assessment of the Evidence for Economy-Wide Energy Savings from Improved Energy Efficiency*, Sussex Energy Group for the UK Energy Research Centre's Technology and Policy Assessment, UKRC, London.

Sorrell, S., Dimitropoulos, J., & Sommerville, M. (2009) Empirical estimates of the direct rebound effect: A review, *Energy Policy* **37**(4), 1356–1371.

South Hams District Council (2015) [Online], Available: http://www.southhams.gov.uk/article/4052/Sherford-New-Community [19 Aug 2015].

Turner, C., & Frankel, M. (2008) *Energy Performance of LEED for New Construction Buildings: Final Report*, New Buildings Institute, March 4, 2008 [Online], Available: http://www.usgbc.org/Docs/Archive/General/Docs3930.pdf [19 Aug 2015].

UK Government (2015) Details of the UK New Towns Act 1946, [Online], Available: http://www.legislation.gov.uk/ukpga/1946/68/contents/enacted.

Van Raaij, F., & Verhallen, T. (1983) Patterns of residential energy behaviour. *Journal of Economic Psychology* **4**, 85–106.

Yu, Z., Fung, B., Haghighat, F., Yoshino, M., & Morofsky, E. (2011). A systematic procedure to study the influence of occupant behavior on building energy consumption. *Energy and Buildings* **43**, 1409–1417.

Zero Carbon Hub (2014) *Closing the gap between design & as-built performance: Evidence Review Report*, Zero Carbon Hub, London. [Online], Available: http://www.zerocarbonhub.org/sites/default/files/resources/reports/Closing_the_Gap_Between_Design_and_As-Built_Performance-Evidence_Review_Report_0.pdf [4 Apr 2014].

8. The practicalities of building with sustainable technologies

8.1 Building with sustainable technologies and innovations

Many of the previous chapters have been devoted to the design, assessment, and construction-related factors associated with the production of sustainable buildings and structures. Many of the latest versions of the technologies highlighted in this text are relatively new to many of the construction markets across the globe, wherein lies an opportunity but also a danger. Designs are often based upon assumptions and these assumptions emanate from previous practice and custom. What if there is very little previous work to rely upon? Few exemplars to gain inspiration from? This situation then tends to increase the risk of inaccurate assumptions, resulting in either poor operational efficiencies or an inoperative technology. If the design does, however, base the fundamental assumptions upon realistic estimates, the technology then needs to be installed following good practice and by trained operatives who understand the best way of getting the most from technological systems. This approach will then minimise any risks that come with most forms of renewable energy sources with all the inbuilt variability that entails. This chapter takes the more theoretical elements of the previous chapters and looks at the interaction between the system design, installation, and maintenance of sustainable technologies.

8.1.1 Defining sustainable technologies and innovations

It has been assumed that once the guidance and advice contained in Chapter 3 relating to the selection and design-related issues of creating a sustainable building have been digested, the decision to use a technology has been either largely made or at least narrowed to a list of leading contenders. Various sources of guidance and help to rank shortlisted

Sustainable Construction Processes: A Resource Text, First Edition. Steve Goodhew.
© 2016 John Wiley & Sons, Ltd. Published 2016 by John Wiley & Sons, Ltd.

sustainable technologies through insights into siting, payback periods, and the general advantages and disadvantages of various technologies are available.

Helpful publications come from a range of sources that include the UK Carbon Trust, whose outputs are often aimed at organisations that wish to assess the appropriateness of renewable energy technologies for an already acquired site. The mix of clear explanations and examples based on case studies in its free downloadable publications provides a relatively unique insight to clients, designers, and especially contractors. Its publications embrace a relatively wide definition of renewable sources, including biomass, digesters, and heat pumps as well as the more narrowly focused wind and solar technologies. This raises a fundamental issue, the aspect of defining what a sustainable technology actually is when related to buildings. Before embarking on a more focused examination of the siting/installation/running/maintenance issues of sustainable technologies (STs) and innovations and defining these, the necessary next step is to look at what is or is not an ST.

This task is not an easy one; an OECD report issued in the mid-1990s titled 'The Global Environment Goods and Services Industry' (OECD 1995), which deals with technologies, some of which would be called 'sustainable technologies' now, states, 'The boundaries of the industry are blurred and they are not easily defined' and refers to 'clean' technologies and processes such as those that 'minimise pollution and material use and clean products which are less environmentally-damaging'.

The perceived evolution of the meaning of technologies previously described as environmental to those now associated with the term 'sustainable' can be linked to the benchmark 1987 Brundtland Report (WCED 1987).

According to the Brundtland Report's sustainable development definition, 'Sustainable technology' can be defined as 'technology that provides for our current needs without sacrificing the ability of future populations to sustain themselves'. The leap from environmental to sustainable had included the element of providing for future needs of a broad but undefined nature.

Other commentators define STs in relation to avoiding unsustainability: 'We will need to attend to our own advancement from unsustainable technological practices' (Carpenter 1995). This underlines the difficulties associated with categorising technologies as STs when definitions beyond those associated with environmental technologies are difficult to pin down.

Therefore, if technologies are to be considered sustainable they should reflect the attributes of sustainability provided below:

* Minimal resource depletion
* Minimal pollution both from manufacture, use and disposal
* Economically viable for the manufacture, distributer, installer and user
* Socially viable with positive impacts on wellbeing and quality of life
* Offer environmental advantages over existing or alternative solutions

All of these are difficult to further prescribe as the building/project, its use, and the overall rationale for using the ST will vary considerably from instance to instance. However, using these attributes as a guide allows the basis for decisions to be established, and more detailed criteria can be formed in each instance.

8.1.2   Choosing sustainable technologies

The use of STs is driven in many countries by the needs of building owners, clients, and occupiers. If any of these stakeholders are not convinced that an ST is a wise incorporation into their new or refurbished building then, unless legislation or regulations demand the use of STs, they are unlikely to be adopted. Whilst different sectors of the construction industry, supplying different clients with different buildings or refurbishments, makes it very difficult to make any universal assertions as to what makes an ST attractive or not, there is some evidence from the retail sector. One hundred retailers (Aberdeen Group 2008, quoted in Dangana et al. 2012) described the top five drivers for green retail (and by implication the use of STs):

1. Competitive advantage
2. Rising cost of energy
3. Need to increase brand value/equity
4. Need for innovation
5. Present/expected compliance mandate

None of these five points could be seen as being completely altruistic, and when many commercial organisations are answerable to stockholders this is unsurprising. The attributes that make an ST attractive can be matched to the drivers that the decision makers feel are most important. Most STs offer reductions in costs, particularly energy costs in use, and these will automatically be registered on balance sheets. The reductions in energy use associated with many commercial and public buildings in Europe will push the performance categorisation shown on the highly visible energy certificates that are required for all EU buildings into the most advantageous categories, reinforcing positive comment upon the respective commercial brand. Further aspects of brand reinforcement can come from the visible aspects of adherence to sustainability-related regulations, such as the use of renewables to fulfil energy-related criteria stipulated in regulations. Innovation and competitive advantage often go together, one reinforcing the other. Innovations that reduce the energy consumption of leased commercial offices can lead to increased rent for the landlord due to lower running costs for the tenant. The balance of capital investment against the energy saved for a refurbishment for a range of commercial buildings has been investigated by a piece of research by Sweets as part of the Investment Property Forum research programme 2011–2015 (IPF 2012). They found the following:

- A 'market standard' refurbishment can improve the Energy Performance Certificate (EPC) rating by at least one grade for most buildings. The improvement is substantial for an industrial/warehouse building (i.e., rating improves from an 'F' to a 'B').

- All offices can be improved by an extra grade for an additional cost of up to 1% of the refurbishment budget, which is potentially viable for most projects.

These very achievable viable improvements in commercial buildings and associated increased rental income are likely to influence decisions to invest in STs. The legislative environment is changing in the UK in relation to commercial and residential buildings that are let out. The UK Government's Energy Act 2011 heralds some possible legislative standards concerning the minimum requirement of EPCs for landlords letting residential and commercial properties from April 2018. This would ensure that these properties would have to have an EPC rating above the bottom two grades, at time of writing 'F' or 'G' (HM Government 2011a).

There are a number of tools available that can assist contractors, designers, and clients to decide the most attractive ST option from a short list prepared through discussions with all relevant interested parties. Some decision systems predate the terminology that describes STs but can be very helpful, including the 'The Seven "C"s for the Successful Transfer and Uptake of Environmentally Sound Technologies' (ETCUEP 2003).

The decision-making processes can involve a range of accepted more generic tools, such as Analytical Process Hierarchy. A structured procedure has been described by Zainab *et al.* (2013), and this paper establishes appropriate decision criteria in two levels of selections for sustainable technologies for existing retail buildings based on the retailers' and contractors' perspectives. The research involved clarifying the decision context; establishing decision objectives; identifying, clustering, and assessing decision criteria; and finally, quantifying the relative significance of the criteria by using Analytical Process Hierarchy (Zainab *et al.* 2013).

8.1.2.1 Planning permission approval

Many renewable and sustainable technologies rely on natural resources of energy to function. This characteristic normally requires that part or all of the physical presence of the technology is placed on the outside of the building being serviced. This is in turn has a visual impact and the potential to alter the outside appearance of the building. This will then, for some countries, bring the fitment into the domain of the planning authorities. In the UK the relevant planning authority should be consulted at an early stage of any project to determine if planning permissions are required. In some circumstances sustainable technologies that are less intrusive but still mounted externally on domestic buildings, such as photovoltaic (PV) arrays, can be installed under the amendments made in the *General Permitted Development Order (GPDO) (further revised 2015)*, or the *Town and Country Planning (General Permitted Development) (Domestic Microgeneration) (Scotland) Amendment Order 2009* (UK Gov 2009). This grants rights to carry out certain limited forms of development on the home without the need to apply for planning permission. However, this may not be the case in national parks, areas of outstanding natural beauty, and conservation areas.

8.1.2.2   Building regulations approval

Similarly, adherence to local building regulations is another requirement. In the UK all installation work in or around occupied structures will be covered by the building regulations, and there are different sets of regulations for England, Wales, Northern Ireland, and Scotland. Whilst all of the regulations are set out and worded slightly differently, they all have the same aims and objectives of ensuring that the buildings that they cover are built and maintained in the safest, most reliable, and most energy efficient way. All the UK-related building regulations apply to the fitment of sustainable technologies, but the most pertinent are the following sections:

Part A, Structure (England and Wales); Section 1, Structure (Scotland); Booklet D (Northern Ireland)

Part B, Fire Safety (England and Wales); Section 2, Fire (Scotland); Part C, Booklet E Fire Safety (Northern Ireland)

Part C, Site Preparation and Resistance to Contaminants and Moisture (England and Wales); Section 3, Environment (Scotland); Booklet C, Resistance to Contaminants and Moisture (Northern Ireland)

Part E, Resistance to Sound (England and Wales); Section 5, Noise (Scotland); Booklet G, Resistance to the Passage of Sound (Northern Ireland)

Part P, Electrical Safety (England and Wales); Section 4, Safety (Scotland)

England and Wales: HM Government (2015). All parts have their own dated amendments and separate English and Welsh specific editions are available from the planning portal website: http://www.planningportal.gov.uk/buildingregulations/approveddocuments/.

Scotland: Scottish Parliament (2015). Each technical handbook as separate dated amendments and an overarching Technical Handbook Summary Guide are available with the updates for each of the separate handbooks: http://www.gov.scot/Topics/Built-Environment/Building/Building-stand ards/publications/pubtech.

Northern Ireland: Building Control NI (2015). The technical booklets for 2012 are available from the Northern Island Building Control website: http://www.buildingcontrol-ni.com/regulations/technical-booklets.

Once a manageable number of technologies have been selected, there are software tools that help narrow down the selection further. The UK's Chartered Institution of Building Service Engineering offer a free download for their TM38 tool, which uses a spreadsheet format. The tool is offered in associated with the TM38 publication *Renewable Energy Sources for Buildings* (CIBSE 2006). The spreadsheet is a simple tool that will give general guidance for a series of defined building use types that are placed in suburban, semi-rural, or rural locations under severe, normal, or sheltered levels of exposure. The software is produced using general principles, and any situation that varies from the norm should be researched further before any decisions are made.

Where considerable expenditure and multiple roll-outs of STs are contemplated, a feasibility study can be justified. Depending on the possible investment or the size of deployment, the feasibility study could vary in scope from a pilot study using a small deployment that may indicate the performance of the chosen technology, to a desk-based study. The cost benefit analysis needs to be undertaken in the widest sense, taking into account the physical constraints, the linkages to existing energy systems, whether the technology is intended for one user or multiple users, and as described in Chapter 3 the likely demand profiles. Once the STs have been selected and feasibility study(ies) undertaken, the technologies need to be specified, ordered, installed, commissioned, maintained, and at end of life, reused or recycled.

8.1.3 Factors influencing the operation of the more commonly used sustainable technologies

As a precursor to discussing the more construction-related issues around sustainable technologies, the following table outlines the basic features associated with the operation of commonly used systems to power, heat, and supply buildings. Table 8.1 describes the factors that impact the assessment and subsequent choice of sustainable technologies.

8.2 General operational issues

The effective care of any operational issues related to sustainable technologies, as for more traditional technologies, can often hinge upon the handover between contractor and operator/owner. The integration of an appropriate commissioning period is also important for any system that has a variable output due to reliance on renewables, non-standard operating conditions, or other special needs. Many sustainable technologies can take a complete year before the components are subjected to the extremes of operating conditions. This will need a longer-lived relationship than is traditionally common and possibly longer handover periods relating to more comprehensive care packages. This can start at initial handover where building logbooks and information will need to be more comprehensive than for buildings using more commonplace technologies that are more recognisable to operatives, and that can be called out using the usual telephone and internet listings.

The next few sections introduce the practicalities, the installation, commissioning, and longer-term maintenance issues of a series of renewable generations and sustainable technologies: solar PV, solar thermal, wind generation, biomass and ST, rainwater and greywater collection systems, micro CHP, and ground/water/air source heat pumps.

Table 8.1 Sustainable technologies assessment

Factor	Sustainable Technology					
	Solar PV	Solar HW	Wind Generation	Ground/Water/Air Source Heat Pumps	Micro CHP	Rainwater and Greywater Collection Systems
Efficiency	Typically 15% with future promises of 30%+	Conversion varies but can be as much as 75%	Relatively high conversion, but large blade area needed to obtain large powers	Ideal conversion factor (COP) of 4.0 (for each kW input 4.0 output) is viewed as good	Combined heat and electrical energy can reach into the high 90 per cents	200 gal tank can replace 100 toilet flushes
Focus of the designers	To assess the electrical usage, the degree of shading, anticipated values of export/import, and design the size of the array accordingly	To anticipate the hot water usage and design the size of the array accordingly	To undertake an appropriate survey to ascertain whether the generator will be effective at that particular development	To ensure that the output of the heat pump is matched to the end use of the heat and the ground/air/water source is utilised efficiently	To balance the heat and electrical requirements of the building with the size and timings of operation and occupancy patterns	That the rainwater system has suitable uses compared to the capacity of the rainwater storage and if potable water is required that suitable filtration and cleaning systems are built in
Focus of any operatives	The integration of the output of the PV cells with the mains connections; particular attention to minimising the possibility of leakage from roof-mounted PV arrays	The fixing of panels to withstand wind and other loadings and ensuring that fixings that penetrate roofs do not cause ingress of moisture	That the generator is placed in the most likely position to maintain high electrical outputs. To ensure that the installation conforms to any planning issues highlighted by the local authority.	That the installation of any exterior pipework systems associated with ground source elements are in good contact with the ground and the loop or borehole has enough length that conforms to the design requirements	That any grid connections are suitably protected and any hot water connections allow for a sink in summer months.	That the rainwater storage tank is positioned to allow effective overflow in conditions of high rainfall. That any filtration or cleaning facilities are effectively working and cannot discharge non-potable water into water pipes that require water of drinking quality

Focus of owner	That appropriate maintenance cycles are kept up, including cleaning (if required) and yearly checks on any inverters or associated equipment	That appropriate maintenance cycles are kept up, including cleaning (if required). Associated thermal stores and system performance should be monitored to ensure that the size of array does match the installed thermal storage.	To monitor the electrical production of the generator in comparison with its rated output. If large discrepancies occur, then appropriate contractors should be contacted to investigate underperformance.	The system's COP can be monitored and if the system consistently performs below the design output and the COP is dramatically lower than would be expected, a specialist should be contacted.	To monitor the heat and electrical outputs through the provided control systems and the consumption through utility/energy monitors with a keen eye on unusual variances in output that will show up in usage patterns	The building owner needs to be able to effectively check the system for appropriate function with particular focus on the effective supply of water in periods of low rainfall and to oversee the filtration system, reducing the likelihood of blockages or suboptimal operation.
Cost/payback (*Please note that as energy/water prices vary and as the technologies involved with sustainable technologies combine with the economies of larger-scale production, payback periods are likely to reduce. The payback will be linked to the usage and positioning of the building, so any of the figures given have to be viewed in the light of this.*)	Although the costs associated with PV arrays are dropping and their efficiencies are promising to increase, payback for UK regions can often be in the region of 15–18 years (lower with subsidies).	Solar thermal systems fare better than PV arrays in the UK and generally can payback in 7–10 years (lower with subsidies).	Wind generation, more than most sustainable technologies, tends to be more dependent upon the siting and availability of constant high winds. For this reason the payback can vary from around 10 to 30 years.	The payback of heat pump systems rely on the appropriate sizing and design of the installed system. A water source system can with ideal conditions and offsetting both cooling and heating, give a payback as soon as 5 years (lower with subsidies).	CHP systems are generally a safe bet regarding payback periods, as the amount of uncertainty is limited. Given a period of good operation, paybacks of around 7–10 years are relatively common (lower with subsidies).	Rainwater/greywater collection systems' viability and payback rely on the usual balance between capital cost and offset from the water saved and therefore not paid for over the lifetime of the equipment. For any system to have a cost rather than environmentally based payback, the property will have to be using a water meter and the payback will be likely be in the region of 15–20 years (Alan Fewkes, personal communication 2011).

8.3   Solar systems and technologies

8.3.1   Basic knowledge

There are some basic requirements concerning the siting of any units and suitability of solar systems. The following aspects will impact the amount of heat or electrical energy produced from any PV or solar thermal array and need to be taken into account by contractors alongside the guidance given by designers. The following aspects of the practicalities of installing any PV or solar thermal array will therefore often require further optimisation.

8.3.1.1   Sizing

The sizing of the system will normally be undertaken by a building services specialist taking into account the constraints imposed by the building form and use, the site, and the wishes of the client. Many of these have been discussed in Chapter 3. There are some specific publications that can guide the reader. These include *Heating Systems in Buildings—Method for Calculation of System Energy Requirements and System Efficiencies, Parts 1–6: Heat Generation Systems, Photovoltaic Systems*, European Standard EN 15316-4-6:2007 (BSI 2007).

8.3.1.2   Installation sequence

The sequence of installation is a vital element of the successful installation of solar arrays in buildings. The installation into new buildings is more straightforward, depending on the structural and servicing considerations of the design. Retrofitted systems obviously need to take into account the results of careful surveys of existing buildings, but if they are stand-alone from the buildings, they are less reliant on existing restrictions.

The use of 'plug-in' modules can incorporate the major elements of different solar systems and improve the efficiency of installation. Examples of these can include panel/tubes with supports, storage systems, safety systems, and for PVs, grid connections including the system are needed to isolate the PV array when an outage occurs on the grid (known as islanding). (One would not want to shock a maintenance person working on the grid, if they wrongly assume that all power is not connected.) The offsite production of such modules can ensure that the quality of the build and the adjustment of any instances of poor quality can be controlled to a greater extent than through more variable on-site processes.

8.3.1.3   Building-integrated new build

Any new building has a structural layout and associated details that can be easily located from drawings, allowing any ST to be designed efficiently. Any wiring/cabling/fixings can be placed as the building is built and do not require extensive investigations or any making good after interventions. Solar thermal panels or PV tiles that are part of a building's roof or façade offer the dual advantage of supplying energy but also forming part of the building's outer structure. Integrated systems obviously have more complex requirements than a concrete roof tile or an insulated cladding panel. These systems can get hot—they may need ventilation—and dealing with these issues often requires some liaison with the supplier of such systems alongside careful coordination with specialist installers.

8.3.1.4 Retrofit

Any existing building, alongside neighbouring sites, will place restrictions on the placing and installation of any PV of solar thermal array. As described earlier, planning permission may have to be sought. Whilst this is normally an issue for the design team, the contractor may have to adhere to certain stipulations pertinent to the installation and commissioning phases. As with many retrofit projects, the true extent of how much construction work is required on an existing property can be clearer once the works have begun. Issues such as weather protection, party wall issues, service connections and service runs, structural issues, and the adequacy of existing services to cope with the demands of solar generation may mean extra works are required. This can have implications leading to budget issues or compromises related to the extent of an envisaged solar system. However, if a client is trying to reduce the carbon emissions from an existing property that might be considered 'hard-to-treat' (difficult to upgrade in a cost-effective and non-disruptive manner), the use of renewables, particularly solar generation, might be one method of reaching that goal.

8.3.1.5 Operational issues

Solar systems have obvious seasonal variations associated with their output and, therefore, operation. Many of the extensive and increasingly sophisticated control, management, and communication systems can allow building operators/users to rely on such systems to squeeze the most out of their solar array. As such systems often pair complexity with capacity, ensuring the control system and the other ancillary components are fully operational is vital before the system is handed over. An appropriate briefing, both face to face and written/electronic for existing and future staff/owners is needed. The 'what if?' scenarios, likely to be integrated with the essential dos and don'ts beyond the standard FAQs, will need to be included in documentation that will allow 'normal' people to use it, not just specialists. Section 8.8 within this chapter has some additional insights into controlling STs.

8.3.1.6 Maintenance

One aspect common to all solar generation technologies is the relationship with shading and cleanliness. The maintenance of overshading vegetation and a regular cleaning regime are needed for any array to work efficiently. The following sections look further at this.

8.3.2 PV Specific Issues

8.3.2.1 Basic knowledge

As described in Chapter 3, photovoltaic panels use a range of similar materials to convert the sun's rays to direct-current electricity. The modules that make up a PV array must conform to the following international standards: for crystalline types IEC 61215 (IEC 2005), for thin film types IEC 61646 (ICE 2008), for a PV module safety qualification IEC 61730 (ICE 2004), and in Europe modules must also carry a CE mark. A typical installation of PVs is shown in Figure 8.1.

A number of more technical issues related to the installation of PV arrays, including the earthing, lightning and surge protection, cable and inverter sizing, metering, blocking diodes, string isolation, and switching devices, are discussed in detail in the DECC Guide (DECC 2013).

Once the pitch of the array has been determined and, assuming that a roof-mounted solution has been selected, the structure of the roof and the respective PV frame and fixings will need to be checked to ensure that they can withstand the calculated wind uplift. There are a number of methods of verifying these structural considerations, including manufacturer's data, Eurocode 5 (focused on the design of timber structures), and test data related to the components used to support the array (BSI 1995).

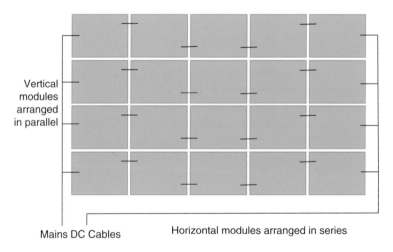

Figure 8.1 A PV system of four parallel connected strings, with each formed of five series connected modules.

8.3.2.2 Building-integrated PV (BIPV) systems

BIPV systems describe a range of PV technologies that can be integrated into or around a building. These technologies replace a more conventional element of a building and undertake its functional requirements as well as operate as part of a PV system. A variety of technologies can be used in the panels, including crystalline silicon and thin film modules. These can be placed onto a building as part of a façade or integrated into a roof as opposed to a series of panels that are placed on a stand-alone frame above a roof. The generating element can be joined so that the overall effect is opaque, such as PV tiles or smaller generating elements encapsulated into glazing, allowing daylight to penetrate a building or space. Walkways and other forms of exterior landscaping and structures can include an element of PV generation, such as shopping centre protection or airport shading. See Figure 8.2 for an example of BIPV.

Hammond and colleagues (2012) established that the energy generated by a 2.1 kWp (peak) BIPV system located in the south of the UK paid back its embodied energy in just 4.5 years. This investigation also undertook a financial analysis from the householder's perspective as well as a cost-benefit analysis from a societal perspective. The results of both indicated that the systems are unlikely to pay back their investment over the 25-year lifetime. This unfavourable financial analysis has to be put into context as the BIPV system would be subject to higher values of insolation in a location with more recorded sunshine (see Table 3.17 in Chapter 3 to compare the relative values of energy from the sun over a typical year in different parts of the world). Many commentators believe that once the cost effectiveness of BIPV systems is improved, the advantages that

Figure 8.2 Photovoltaics in buildings (Photo by Steve Goodhew).

replacing façade and roofing materials offers, as well as the linked generation opportunities, will enable BIPV to become one of the most commonly used STs.

8.3.2.3 Temperature-related effects

The output of PV modules is directly influenced by their temperature. If a crystalline module increases in temperature by 1°C, it is likely that output will decrease by 0.5%. It is therefore wise to allow for appropriate ventilation behind an array for cooling, typically at least a 25–50 mm vented air gap at the back of the modules. For building integrated systems, this is usually addressed by providing a vented air space behind the modules. On a conventional pitched roof, batten cavity ventilation is normally achieved in the UK for pitched tiled roofs by the use of counter battens over the roof membrane and by the installation of eaves and ridge ventilation.

Other more 'technical' factors that can influence the performance of a PV array:

- The operating and electrical characteristics of the chosen panels
- Dirt on the array, especially from nearby trees
- The efficiency of the inverter
- The match between the array and the inverter
- Losses through the cable runs
- Matching of demand with array supply

8.3.2.4 Stand-alone or grid-connected systems?

Stand-alone PV systems require another system to store the generated electrical energy for use outside the array's peaks of performance. Whilst hydrogen and hot water storage systems can be used, they tend to require intermediate steps to establish an electrical supply and do not offer an electrical supply in the case of using the electricity to heat water held in highly insulated tanks. Battery storage, whilst currently not very efficient, low maintenance or inexpensive, is a tried and tested solution to storing generated direct current. Batteries, particularly the more traditional lead acid variants, need careful housing, including access, ventilation, and allowances for connection. Batteries can be connected in series or parallel; series might give theoretical advantages in performance, but practical considerations may lead to parallel configurations. If wired into a parallel circuit it is generally not advised for banks of more than four to be used. Whichever circuit connection is chosen, the batteries need to be fit for purpose, normally deep cycle, allow for good performance even after being discharged, leave a small amount of charge, and be able to cycle over a great many occasions before requiring replacement. Do the batteries need to be vented or sealed? Much will depend on the ability to ventilate the battery storage location due to aspects of the charging operation.

This is particularly important in the case of vented lead acid batteries as the lighter-than-air gas hydrogen is given off during charging and a concentration of more than 4% could explode if ignited. Thus, battery banks must be housed in accordance with BS EN 50272-1:2010 (BSI 2011) or equivalent international standards. Do the numbers of batteries and the charging system have an adequate storage capacity for the PV array? Normally, this is rated by the probable number of days that the storage can supply the predicted demand without further PV electrical input. This battery storage capacity would normally be sized so that the output of the PV array falls between the manufacturer's maximum and minimum recommended charge rates.

8.3.2.5 Inverters

Inverters are electronic devices that take direct current (DC) for a PV array (or any DC source) and convert this to alternating current (AC). As there is an energy loss in this process, some recent building services designs are trying to use DC appliances to increase the efficiency of use of generated energy by avoiding the use of inverters. This is a logical step if the system is safe and a clean supply of DC can be guaranteed for the majority of the time when electrical energy is demanded. These DC systems are still being developed, and issues related to using grid-supplied AC energy and the losses that occur when conditioning for DC use are still being researched.

Using the correct number and size of inverters is vital for the maximum electrical power to be used or exported to the electrical grid. Inverter matching needs to follow the guidance from the inverter manufacturer and make use of any sizing software supplied by the manufacturer. Inverters normally would have to cope with the peak output of an array; however, in certain countries, such as the UK, solar arrays don't normally reach that level. It is therefore common in the UK to size to between 100% and 80% of the array's peak output.

If part of an array becomes overshaded by a passing cloud and the generated power drops below a set value, the inverter will not be able to take any electrical power from the rest of the array, even it is still in sunshine. Having multiple inverters that can take the energy from parts of an array will allow the parts in full sun to still generate whilst the overshaded elements are unable. Inverters used in the EU must have the ability to prevent generation when the rest of the grid is not receiving electrical power. This effect, known as islanding, is prevented by the inverter carrying a Type Test certificate to the requirements of Engineering Recommendation G83/1 or G59/2 (ENA 2011).

8.3.2.6 Hydrogen storage

An alternative storage option is to use the PV array to generate the electricity that will then be used to produce hydrogen gas via an electrolyser that can be stored for later use to power a fuel cell that in turn will produce

> ### 8.1 Want to know more? PV installations
>
> Helpful publications/resources:
>
> BS 7671:2008 Requirements for electrical installations (all parts—but in particular Part 7-712, Requirements for special installations or locations—Solar photovoltaic power supply systems) (BSI 2008).
>
> BS EN 62446:2009 Grid connected photovoltaic systems: Minimum requirements for system documentation, commissioning tests and inspection (BSI 2009a).
>
> BS EN 50272-1:2010 Safety requirements for secondary batteries and battery installations. General safety information (BSI 2010a).
>
> Paul Blackmore's BRE text: P. Building-Mounted Micro-Wind Turbines on High-Rise and Commercial Buildings (FB 22), BRE Press.

electrical power. In common with the use of batteries, storing energy in a different form from the original energy source requires energy conversion. Energy is lost at each point of conversion. This is offset by the ability to use renewable energy at any time the building user requires. As with batteries, space, safety, and knowledgeable husbandry are required to allow hydrogen storage to be legitimately incorporated into a building's energy strategy.

8.3.3   Solar hot water specific issues

8.3.3.1   Basic knowledge

The purpose of solar thermal systems, as described in Chapter 3, is to reduce the amount of energy used by a building to heat or preheat air or water used in a building. (As a point of principle, when attempting to generate hot water using renewables or STs it is wise to first reduce the quantity of hot water required so a minimum amount is wasted before looking at system requirements).

Depending on the system used; the locations, orientation, and demand for hot water/heated air; assumed discount rates; and any savings on maintaining an existing water/air heating system, the payback period in relation to a solar thermal system can commonly vary between approximately 7 and 12 years. This can be reduced through subsidies that are available in some countries. Operational outputs from solar thermal systems have been discussed in Chapter 3, and evidence can be viewed in the UK Energy Saving Trust's field trial of solar thermal systems (EST 2011).

Solar thermal or hot water systems absorb the radiant heat energy from the sun's rays and transfer those to a fluid (liquid or gas), which then transports this heat energy to a heat exchanger. This exchanger in turn transfers this heat energy to systems and building services that require that heat. In most solar thermal installations the systems are a combination of closed and open elements. The circuit that might be termed the primary circuit runs from the absorber to a heat exchanger within the building, circulating in a closed loop. The secondary circuit takes that heat energy and supplies the heat to the point of consumption, and this is an open circuit, the fluid or water needing to be replenished if it is used for heating spaces or hot water (Figure 8.3).

As for PV systems, there are elements of common practice that can quickly be used to check whether an installed system will function as the design team envisaged. The solar thermal system needs to have a large enough collector area to allow worthwhile rates of heat collection, but the size of the collector area also needs to be balanced with the rest of the system, including necessary storage capacity (if included). A commonly sized collector for a domestic system is approximately 4 m², but commercial systems can be considerably larger. The output from a 4 m² domestic system will vary depending on whether an evacuated tube or flat-plate collector has been selected. Rates of heat collection of approximately 450 kWh/m² or more per annum for flat-plate collectors and 550 kWh/m² per year for evacuated tubes are common in the UK. This will result in annual carbon savings of around 100–110 $kgCO_2$/m² for flat-plate collectors and 130–140 $kgCO_2$/m² for evacuated tubes when compared with an equivalent gas boiler supplying hot water (Thorne 2011). If the building being serviced by the solar hot water system is situated in a location where the

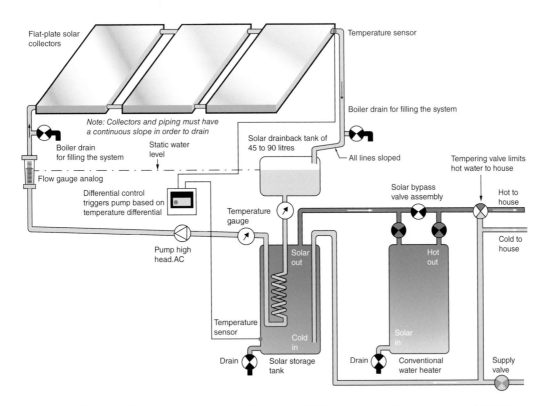

Figure 8.3 Diagram of solar thermal systems (Pennycook 2008, reproduced by permission of BSRIA).

risk of breakage is higher than normal, whilst the higher potential output of the potentially more delicate but more efficient evacuated tubes might be tempting, the more robust but less efficient flat-plate collectors might be substituted.

To reduce the loss of absorbed heat in the primary circuit (the pipework that links the collectors to a heat exchanger or storage vessel), the system needs to be insulated, allowing the greatest percentage of the heat that has been absorbed to be delivered. Weather-resistant external pipe insulation is essential, particularly in relation to resisting moisture effects and UV degradation. More conventional insulation, both for pipework and storage vessels, can be used on the normally internal hot water storage cylinder and connecting pipes.

Whilst appearing obvious and covered in Chapter 3, the collectors need to be placed so that they ideally face close to south (but southeast and southwest orientations are generally next best), followed by laying collectors at a tilt of approximately 40 degrees, (depending on the county of use). If the shading situation varies from that assumed in the system design—either through new building works, local vegetation, or other reasons—will this impact the system's performance? A shade-free system is preferred, but reasonable performance across the year can be expected if it is south facing and partially shaded by, for example, a dormer or protruding roof light or deciduous trees.

The effective use of solar thermal systems often rely upon the behaviour of the occupants of the building using the system (as can be seen in Chapter 6). Many solar thermal systems are not particularly intrusive and when roof mounted can be almost invisible from the ground. A legitimate question (although it might seem rather strange) is, 'Do the occupants realise that they have a solar supply of hot water or air'? and 'Do they realise that there are optimum behaviours that will allow them to maximise the impact of this resource'? (for instance, by using the hot water in the morning and allowing the system to absorb heat and preheat or top up the intended target for that heat during the day time). If the solar thermal system has a display that monitors and shows the performance of the system, is it easily understood? Do the occupants realise the impact of what the display means? And can the occupants/users/maintenance personnel have access to the monitoring system, which will inform them if the system is broken or working inefficiently? Will the display show if any aspect of the use of heating or hot water will affect the use of backup or auxiliary heating

---

8.2 Want to know more? British and European Standards for solar collectors, solar water heating systems

BS EN 12975:2006 Thermal solar systems and components (BSI 2006a).
  BS EN 12976:2006 Thermal solar systems and components. Factory made systems (BSI 2006b).
  DD CEN/TS 12977-1:2010 Thermal solar systems and components. Custom built systems. General requirements for solar water heaters and combi systems (BSI 2010b).

systems and their efficiency? Much of the potential of any system relies upon the initial commissioning. Appropriate commissioning of the system over a time period is advised by the manufacturer/specialist installer to ensure that no minor issues with pipe or ductwork alongside air pockets or fan/valve directionality might be reducing the impact of the solar collection.

8.4 Wind generation

Many of the characteristics of wind generation systems echo the issues from PV systems. If the wind generation system is installed without taking into account fundamental principles, it is likely that the assumed average electrical output will not be reached. As one of the prerequisites of solar systems is access to direct sunlight, a wind generator will not give assumed outputs if the installation does not allow full access to the highest average wind velocities used in any calculations. The larger wind generators tend to require masts of enough height allied to suitable positioning to ensure the blades or other wind-catching device have a maximum airflow over them. Smaller generators can be roof mounted, but according to a BRE report (Blackmore 2010), great care in the siting of any wind generator mounted upon a building is necessary. This section focuses on the issues related to building-mounted wind generators, some of which also apply to larger turbines.

Theoretically, the power P produced from a turbine can be described by the following:

$$P = v^3 A \rho$$

where v is the wind speed, A is the swept area (if the turbine is a horizontal axis machine, where r is the radius), and ρ is the density of air.

This relationship of extracted power to velocity, air density, and area of blade has implications upon how the power from a turbine can be maximised. If the radius of a horizontal axis turbine, commonly used for building mounted applications, is increased by half as much again (in this instance let us assume this doubles the swept area), this should theoretically double the power output. If the wind speed is doubled, the power output of the original turbine would be increased eightfold. This would point to decision makers ensuring that the turbine is placed to allow the average wind speeds flowing over the blades to be maximised as well as taking into account the size of the turbine itself. This statement needs to be qualified, however, as the devices within most commercially available turbines restrict the turbine speed for reasons of mechanical preservation. Therefore, there is a maximum amount of power that can be extracted from a wind turbine (situated in non-turbulent air; a free stream). This can be estimated through the use of a constant, 16/27 (0.593), or the maximum output from a turbine will be no more than approximately 59%, which is known as the Betz limit.

This will then limit the power output according to $0.593 \times \dfrac{(v^3 A \rho)}{2}$ (Carbon Trust 2008).

Whilst this is a theoretical estimation of power output from a wind turbine, a more realistic and relatively commonly used method is a load factor. This takes the measured output of an installed turbine in a specific location and compares this measured value with the maximum rated output. An alternative method is to assess the annual energy production yield, a method that is part of the British Wind Energy Association (now Renewable UK, http://www.renewableuk.com) standard used to establish the reference annual energy generation of a small-scale wind turbine. This uses an average wind speed of 5 m/s for this assessment and estimates the turbine's annual yield of energy. The Energy Saving Trust's domestic small-scale wind field trial report (EST 2009) used load factors to show the outputs of measured electrical power from 57 domestic installations for one year ending in March 2009. The sample had installations located across the UK and Ireland covering nine turbine models and a range of building and nearby freestanding mast-mounted systems. The findings of the report were that there were distinct differences between the load factors of building-mounted and freestanding turbines:

> Free standing turbines installed in the appropriate location with an undisturbed wind resource were seen to have very good performance with annual load factors in some instances in excess of 30 per cent.

Whilst building mounted turbines exhibited generally poor output due to installations at sites with inadequate wind speeds, those in rural or exposed locations achieved load factors in excess of 5 per cent. The best building mounted site in the field trial was located in Scotland with a load factor of 7.4 per cent (Reproduced by permission of EST).

The lessons from the field trial are obviously limited in scope, as this is meant for the UK and only for small installations. However, the themes that can be observed are as follows:

- Freestanding turbines in locations of high average wind speeds sited away from obstructions, such as other buildings, will have the best chance of generating enough electrical power to warrant their installation.
- The far north of the UK, with special attention paid to Scotland, would be best suited for wind turbine locations.
- Using an anemometer to assess the wind speed for as long a time as is feasible (one calendar year would be good) is a wise precursor to deciding to locate a wind turbine in a specific site that has unique obstructions or topography around it.

If a suitable site has been found, a desk study can commence, looking at the feasibility of locating a wind turbine, freestanding or building mounted, on that site. A first step is normally to assess the wind potential via historical records. There are a number of databases for the UK including the BERR wind speed database, which is now archived (HM Government 2009), and the wind data for a particular region can be found using a grid reference or postcode. Very crudely, if the average annual wind speed is less than 6 m/s, it is unlikely that the installation of a wind turbine will be beneficial.

8.4.1   Noise from turbines

Whilst the focus of this section is on the smaller scale turbines that might be included in the generation strategy of a sustainable building, noise can be an issue with small generators as well as the often more publicised perceptions of large-scale onshore wind turbines.

Noise from wind turbines normally comes from two possible sources, aerodynamic and mechanical (Lowson 1993; Lowson & Fiddes 1994).

According to the ETSU-R-97 (1996) report, aerodynamic and mechanical noise is emitted by a wind turbine through a number of sources:

- self-noise due to the interaction of the turbulent boundary layer with the blade trailing edge
- noise due to inflow turbulence (turbulence in the wind interacting with the blades)
- discrete frequency noise due to trailing edge thickness
- discrete frequency noise due to laminar boundary layer instabilities (unstable flow close to the surface of the blade)
- noise generated by the rotor tips

Mechanical noise can come from these sources:

- the gearbox and the tooth mesh frequencies of the step-up stages
- generator noise caused by coil flexure of the generator windings, which is associated with power regulation and control
- generator noise caused by cooling fans
- control equipment noise caused by hydraulic compressors for pitch regulation and yaw control

To measure the sound emitted from a small turbine, the standard IEC 61400-11 *Wind turbines—Part 11: Acoustic noise measurement techniques* (IEC 2012) has an appendix containing details of the procedures and equipment needed. Some of the methods and complexities are not easily understood, so there is a small wind turbine annex F which keeps methods simpler and relevant for small wind turbines. Using the standard and an appropriately qualified consultant, a noise map can be generated for a turbine that explores the sound pressure level measured in dB(A) at various wind speeds. The maps will often use a traffic light coding system to alert to red/amber/green areas, which inform users whether the noise from the turbine is deemed intrusive or not. Typically, the associated sound pressure levels might include the following classification:

- Red: predicted noise level greater than 45 dB(A)
- Amber: predicted noise level between 40 and 45 dB(A)
- Green: predicted noise level below 40 dB(A)

8.4.2   Maintenance

Wind turbines vary in design and therefore can have varying maintenance requirements. Most conventional turbines have an expected lifetime of between 20 and 25 years; however, some reports (Mackay 2012) have estimated economic lifetimes of between 12 and 15 years even if the turbine could have a longer operation window.

The predicted yearly cycle of use, and therefore the maintenance of small-scale wind turbines, is difficult to estimate as wind turbines have only been produced in large numbers for about a decade, and technologies that underpin the design of those turbines entering service are different in subtle ways from those of 10 years ago. Requesting service-related data from potential turbine suppliers/installers would be appropriate. In relation to much larger turbines, it can be assumed that the total annual charges represent a percentage of the installed cost, often quoted between 3% and 5% (Milborrow 2010). Some smaller systems that use battery storage have additional maintenance issues, including a battery life that is likely to be in the region of 6–10 years.

8.5   Biomass

As described in Chapter 3, the term *biomass* can refer to many forms of short-time rotational plants, as well as biofuels such as biodiesel and biogas from digesters. As the drivers for the use of such fuels include the local availability of waste or semi-waste products, the location of any facility is very important. This is twofold, not just the proximity of the fuel but the predicted continuity of supply of that fuel. Table 8.2 shows the range of wood/plant/animal-related fuel types, and each of these can suit buildings in different locations.

Depending on the chosen fuel, there are three common methods of obtaining energy from biomass: direct combustion, anaerobic digestion, and gasification.

8.5.1   Direct combustion

This is probably the most appropriate method for application in buildings. This can vary from a small domestic wood-burner consuming logs to a district heating system using large quantities of timber-based pellets. The calorific value and other properties of the different timber and plant-based fuels can be seen in Table 8.3.

Table 8.3 shows that the energy density (with the exception of wetter woodchips) is relatively similar across the different fuel feedstocks. However, because the bulk density varies between these fuels, the energy density is much more variable and the amount of storage needed to obtain the same energy does vary considerably. The volume of the grass *Miscanthus* might be five to six times greater than wood pellets to supply the same energy.

Table 8.2 The categories of materials that make up the different options for using biomass

Category of Biomass	Separate Subcategories
Timber-based or virgin wood	Logs, brash (wood that has been pruned from trees on verges of roads and railways and similar materials from parks), bark, wood chips, (chipped pieces of timber often 20–50 mm in diameter), wood pellets (manufactured from sawdust compressed and forced through a former or die to produce particular diameter cylinders), sawdust
Crops grown for energy use	Short rotation energy crops (such as poplar and willow on a 2–5 year cycle, shorter than traditional forestry but longer than non-woody energy crops) Grasses and non-woody energy crops *Miscanthus* (elephant grass), switchgrass, reed canary grass, rye, giant reeds, agricultural energy crops (oil seed and sugar crops)
Waste or residues from agriculture and food wastes	Straw, rice husks, poultry waste, manure, slurry, silage, and domestic, commercial, and industrial food wastes all converted to heat by combustion or to flammable gas by anaerobic digestion
Liquid biofuel (sometimes a subcategory of the above categories, but becoming commonly used to power combined heat and power systems, so worth identifying individually)	Bioethanol, biodiesel (from waste oil or from energy crops), pure vegetable oil

Table 8.3 Typical properties of timber- and plant-based fuels related to use in biofuel energy systems for buildings

Fuel	Energy Density GJ/tonne	Energy Density by Mass kWh/kg	Bulk Density kg/m³	Energy Density by Volume mJ/m³	Energy Density by Volume kWh/m³
Woodchips (moisture dependent)	7–15	2–4	175–350	2000–3600	600–1000
Logs (air dry)	15	4.2	300–550	4500–8300	1300–2300
Wood (oven dry)	18–21	5–5.8	450–800	8100–16800	2300–4600
Wood pellets	18	5	600–700	10800–12600	3000–3500
Miscanthus (bale)	17	4.7	120–160	2000–2700	560–750

Source: BSRIA Illustrated Guide to Renewable Technologies by Kevin Pennycook (2008). Reproduced by permission of BSRIA.

This would need to be taken into account in the system design and operation. The moisture contents also have a marked influence, both for the calorific value and upon the operation of supply plant, wetter stock possibly introducing more blockages than drier fuel.

Combustion of biomass works best when the boiler in question is working at full capacity. Some systems can allow for lower capacity working or modulation; however, this often does not provide the optimum combustion environment for full efficiency. One method of allowing the boiler to work at its best is to link the system to a well-insulated storage tank, and when the boiler is working any excess hot water can be stored for more intermittent use. This does, however, introduce the expense and space requirements of additional plant and machinery and the accompanying losses of pipes and stage vessels.

Many issues need to be taken into account when integrating a biofuel solution, and the sizing of the boiler and the associated system is one of the most important. The UK Carbon Trust has online guidance and a spreadsheet tool that can help: http://www.carbontrust.com/media/63116/biomass-software-tool-user-manual.pdf (Carbon Trust 2012).

8.5.2   Anaerobic digestion

Anaerobic digestion uses fuels that can be broken down in controlled conditions to supply methane gas and any associated solid and liquid components. The gas can be used to power a turbine, generating electricity and also heat. The solid and liquid by-products can be used (subject to contamination levels) as a soil enhancement. Beyond the tanks required for the digestion, there will be an equipment requirement for the preparation of the fuel stock (such as agricultural wastes, manure), often reducing the individual particle size, which gives a larger surface area for the bacteria to act upon. Storage of the by-products will also be necessary, and some thought should be given to the sensitivity of the fuels and by-products when deciding on solutions in sensitive and highly urbanised sites.

8.5.3   Gasification

Gasification normally entails the use of a fuel that when heated will give off flammable gas in a reactor or vessel. When wood fuel is heated in the presence of a controlled quantity of an oxidiser, methane and other combustion gases are given off. The constituents of the gas will vary depending on the reactor type, the oxidizing medium (air, steam, or oxygen), and the fuel. The reactor types vary in their use of top, side, and bottom feeds and exits for oxidiser, fuel ash, and product gas. The gas produced often can be fed to a gas engine as part of a combined heat and power system.

Most of the above biomass fuels require storage, feed, and sometimes drying systems as part of the package that fuelling with biomass entails. These factors need to be taken into account when contemplating biomass as an energy option against other systems. This aspect does vary between solutions, with anaerobic digestion probably requiring the largest dedicated area for deliveries, tanks, pipework, and so on, as many biomass energy installations feed into combined heat and power units.

As described in Chapter 3, in its most basic sense a heat pump system transfers heat from one location to another using electrical energy to drive the system. Its major basic components include heat exchangers, pipe and/or duct work, a compressor, an expansion valve, pumps and fans, control gear, and often buffer tanks. A basic schematic of a heat pump system is shown in Figure 8.4.

The main reason for using a heat pump is to supply heat through the use of a relatively small amount of electrical power aiming to produce over twice the amount of heat energy. The coefficient of performance (COP) of a heat pump is the ratio of the heat that is obtained to the electrical power that is required to work the heat pump—in effect, a multiplier of one form of energy to another. This can be calculated using the temperature of the condenser (T_{COND}) and the temperature of the evaporator (T_{EVAP}) (both measured in Kelvin) to approximate the heat energy output and the energy put into the pump to run the cycle.

$$COP = \frac{T_{COND}}{T_{COND} - T_{EVAP}}$$

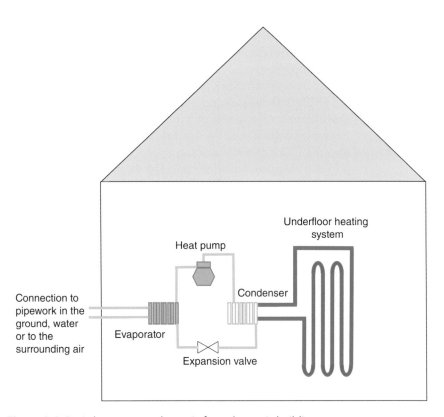

Figure 8.4 Basic heat pump schematic for a domestic building.

Table 8.4 The typical COPs for a water-to-water heat pump associated with different heat distribution systems

Heat Distribution System	COP
Under-floor heating (30°–35°C)	4.0
(35°–40°C)	3.5
(50°–60°C)	2.5

Source: *BSRIA Illustrated Guide to Renewable Technologies* by Kevin Pennycook (2008). Reproduced by permission of BSRIA.

An ideal COP (which in reality is never really achievable) can be compared with the actual COP of a heat pump, which gives a value known as the Carnot efficiency. The Carnot efficiency tends to vary between about 0.3 for normally encountered space conditioning units and 0.7 for the very latest highly efficient electric heat pumps.

One factor that does influence the COP of heat pumps is the final temperature that the heat distribution system will be operating at. The lower that operating temperature is, the more efficient the system is likely to be. This favours underfloor heating or lower temperature systems, as can be seen in Table 8.4.

A European Union project SEPEMO-Build (seasonal performance factor and monitoring for heat pump systems in the building sector) aimed to collect robust data about heat pump performance from case studies that influence both the pump system's reliability and seasonal efficiency. This project was completed in 2012. A summary of its findings can be found at http://sepemo.ehpa.org (SEPEMO 2015).

SEPEMO-Build defined a series of seasonal performance factors (SPFs) that relate to the parts of a heat pump system included inside a system boundary, which may be not just the heat pump and heat exchanger but also the more directly building-related heating or hot water systems. These and the respective performance of 45 systems across Europe were evaluated over a calendar year. The SPFs have a large influence on the COP of a heat pump system and therefore can alter the potential viability (the ratio of electricity input to heat out) of installations. Comparisons between the parts of the system included for the European Standard EN14825 (BSI 2013), the typical standard used for many heat pump evaluations, and the proposed systems boundary from the SEPEMO project can be seen in Table 8.5.

As can be seen, the season factor SPF_{H4} includes more of the other electrical use of parts of a heating/hot-water system than does the European standard, and this is recommended by the UK's Energy Saving Trust (EST 2013) as the factor to use when assessing the COP of a heat pump system. This factor, alongside others described in Chapter 3, influence heat pump operation. The SEPEMO project describes the main parameters influencing systems efficiency:

- the efficiency of the heat pump unit
- the quality of installation

Table 8.5 The parts of a heat pump system included inside the system boundary for different seasonal performance indicators (SEPEMO 2015)

System Component	EN14825	SPF$_{H1}$	SPF$_{H2}$	SPF$_{H3}$	SPF$_{H4}$
Heat supplied by the heat pump	Yes	Yes	Yes	Yes	Yes
Electrical part of the heat pump system					
Compressor	Yes	Yes	Yes	Yes	Yes
Heat source fan/pump (depending on air of fluid heat transfer medium)	Head losses		Yes	Yes	Yes
Back-up system for heating and hot water				Yes	Yes
Buffer tank pump					Yes
Space heating and domestic hot water pumps and fans	Head losses				Yes

Table 8.6 The heat supplied compared to the degree days for a particular site in the UK's EST second-phase field trial (EST 2013, reproduced by permission of EST)

	Heat Output (kWh)	Electricity Input (kWh)	Degree Days	Heat Supplied/ Degree day
April 2009 to Mar 2010	4206	2574	2057	2.04
May 2011 to Feb 2012	3925	7155	1541	2.54

- the design of the system and temperature level of the heating system
- the insulation level of the building envelope, and
- the climatic condition where the heat pump is employed.

Picking up the last of the listed parameters, the climatic conditions, it is possible to benchmark the performance of a heat pump by using degree-days (degree-days are discussed in Chapter 5, which looks at assessment systems relevant to sustainable construction). Degree days give the cumulative number of degrees Celsius that a building would need over one given year to maintain a set interior temperature. Therefore, the larger the number, the more energy would be expected to be used to heat a property. The Energy Saving Trust monitored a range of heat pump systems used in buildings, and Table 8.6 is based upon their second-phase report (EST 2013), described in more detail in the next few pages.

The location of the development, as discussed within Chapter 3, will influence the choice of air, water, or earth as the main medium to extract or deposit heat to and from the heat pump. Each medium has its own unique set of issues associated with its installation. Air source systems have the least problems as far as where the unit takes in and exhausts air.

Water source heat pumps rely on an appropriately sized expanse of water to ensure any coils, heat exchange matrix, or pipe-work need not overheat or freeze. Consideration also needs to be made for the

Table 8.7 The range of efficiencies of air and ground source heat pump systems monitored over one year in the UK (EST 2010, reproduced by permission of EST)

	Heat pump COP		System Efficiency	
Source of heat	Air source	Ground source	Air source	Ground source
Range of efficiency	1.2–3.3	1.3–3.6	1.2–3.2	1.3–3.3

limitation of accidental damage by any water-based activities and essential maintenance.

The choice, and therefore the installation, of the two main types of ground source heat pump systems is dictated by the plot area and sub-strata of the development. Larger plots can use coiled horizontal pipe-work placed in a trench, 0.5–1 m deep with a minimum separation of pipes of approx 0.3 m. Larger heat pumps will require a longer pipe placed in a longer trench.

For plots of limited space, a vertically drilled borehole will be a better solution. The borehole will often be in the region of 15–150 m deep and contain a 100–150 mm diameter pipe flow and return pipe at least 5 m apart leading to and from the heat-pump system. The bore hole needs to be long enough to allow good heat transfer to the surrounding earth, and a portable thermal transfer measurement system can be used to assess the capacity of the borehole and assist in sizing the heat pump to be installed.

A year-long field trial of air and ground source heat pump systems that was carried out by the UK's Energy Saving Trust monitored the technical performance and customer behaviour at 83 sites across the UK (EST 2010). The findings confirmed that there was considerable variance between installations, as can be seen in Table 8.7.

The field study drew a series of conclusions related to the effective installation, commissioning, and running of heat pump systems, which in summary stated (Reproduced by permission of EST):

1. Heat pumps are sensitive to design and commissioning. The field trial covered a variety of early installations, many of which failed to apply the heat pump correctly. This result emphasises the need for improved training.
2. Keep it simple. There were many system configurations monitored in the field trial. In most cases, the simplest designed systems performed with higher efficiencies.
3. The impact of domestic hot water production on system performance is unclear. Heat pumps can be designed to provide domestic hot water at appropriate temperatures, but more investigation is required to determine the factors which impact system efficiency.
4. Heating controls for heat pump installations have to be comprehensively reviewed. There has been a failure to explain proper control requirements to both installers and heat pump customers.
5. Responsibility for the installation should be with one company and ideally be contractually guaranteed to ensure consistency in after-sales service.

This first phase of heat pump trials has recently been supplemented by the findings of a second phase trial published in May 2013 (EST 2013). The report focuses on 38 of the original report's 84 heat pumps, investigating a series of different interventions that were intended to improve the performance of the installed system. The interventions were quite varied and are summarised in Table 8.8.

The typical costs of a commercial system for a larger building, in this case a hotel, are shown below in Table 8.9.

Table 8.8 The different interventions witnessed during the second phase of the EST heat pump field trials (EST 2013, reproduced by permission of EST)

Scale of intervention	Intervention
Major	Changing the pump for a different capacity
Medium	Adding buffer tanks, changing radiators, replacing circulating pumps with more effective types
Minor	Refilling ground loops, changing controls

Table 8.9 Costs incurred when installing a ground source heat pump system to a 8,980-m^2 Greek hotel in 2008 (SEPEMO 2015)

Cost element	Cost (Euros)
4 water wells (60-m depth each one)	30,000
2 water source heat pumps (HAUTEC, Type HWW-PN-294/4)	280,000
Pumps—circulators	20,000
Plate heat exchangers (Ti HAUTEC, Type T50M HV-23-CDS-10)	40,000
Pipes—expansion tank—electricity supply	90,000
Study—installation, electrician, and hydraulic works	32,000

8.3 Want to know more? Installation of heat pumps

Alongside the links and references already cited within this section, further information related to the installation of heat pumps can be obtained from the UK's Carbon Trust:

Ground source:
http://www.carbontrust.com/media/147462/j8057_ctl150_how_to_implement_guide_on_ground_source_heat_pumps_aw__interactive.pdf (Carbon Trust 2015a).

Air source:
http://www.carbontrust.com/media/147466/j8058_ctl151_how_to_implement_guide_on_air_source_heat_pumps_aw.pdf (Carbon Trust 2015b).

8.7   Micro CHP

Micro CHP systems are normally designed as combined heat and power units that are small enough to be used as domestic options. The EU classification actually classifies any system under 50 kW as a micro system. For the purposes of this text it is assumed that micro units are on the whole used in domestic surroundings.

As mentioned in Chapter 3, there are two main engines/cells that are being used for the most readily available domestic micro CHP systems: Stirling engines and fuel cells. These may well be joined by some more varied technologies in the future.

The Stirling engine is a closed-cycle regenerative heat engine (technically an external combustion engine) that is sealed. This sealed aspect allows it to be quieter and less intrusive than the more traditional internal combustion engines used in past CHP systems. The power unit is heavier when compared to an internal combustion engine of an equivalent size and takes heat from the surroundings to expand a gas inside a cylinder which, in turn, exerts a force upon a piston to create rotary motion that will turn an electrical generator. This form of engine is very quiet and therefore very suitable for domestic operation. However, the amount of electrical power generated compared to the heat output is approximately 1:6 in favour of the heat energy (He & Saunders 2011).

Fuel cells can be viewed as a form of continually generating battery, producing electrical power whilst the cell is supplied with fuel (normally hydrogen and oxygen). The first large-scale commercial deployment of a fuel cell as a CHP system in the UK was opened in 2003 in Woking Park (WBC 2007). This installation used a 200 kWe fuel cell (alongside a number of more conventional CHP units running on gas), 9.11 kWp solar energy photovoltaics, heat-fired absorption cooling, and thermal store. The total capacity of the installation is 1.2 MWe and 1.6 MWth, considerably larger than a domestic-scaled micro CHP unit.

Most modern fuel cell units intended for the domestic market have three main components: the fuel cell(s), a heat storage unit, and a control system to manage the heat and electrical energy. As described in Section 3.4.3 of Chapter 3, fuel cells have no moving parts so they involve very little noise or vibration. In that respect, they are very suitable for the domestic market. Recent mass-market versions have been produced by a well-known domestic boiler maker, and these promise to supply up to 100% of the heating and 75% of the electrical requirements of a home. Although mini and micro CHP (co-generation systems) are similar in size, with some of the installation requirements of a conventional boiler, many of the issues concerned with introducing this technology into new or existing buildings are linked with the electrical power output alongside the heating capacity. The electrical supply, storage, and export issues can be virtually eliminated with an installation that has a grid connection as, unless the base load is not exceeded by the electrical generation, all spare electrical power can be redirected to the grid. The usual specialist control systems need to be used for any grid connection to ensure the CHP unit does not supply electrical power to the grid when the network has been shut down.

Summer conditions can present problems for the correct operation of CHP units. Domestic properties only require limited amounts of hot water; however, the CHP unit generates hot water whenever it produces electrical power. Thus, the installation will normally require controls that can divert the hot water to suitable sources within the building. This may entail the enlargement of any existing hot water storage capacity and/or a suitable 'sink' for any excess hot water.

8.8   Control systems and smart home energy management

The ability to control systems often defines or at least influences the performance of many renewable and sustainable technologies. STs make use of natural energy sources and provide comfortable internal environments for building occupants whose varying behaviour can be difficult to satisfy. Internal controls can be time, occupancy, or condition related. Once STs are introduced into buildings, the interactions between patterns of behaviour and patterns of energy supply or 'best times' for efficient generation become beyond the capabilities of traditional on-off/timer-temperature control systems. The constraints that relate to renewables and other methods of onsite generation include the need to link energy demand to energy supply. This implies that an active rather than a passive link is required. A measurement of energy consumption can trigger the activation of a micro CHP plant. The generation of solar or wind related energy can prompt the use of a matched energy use in the building that is flexible enough to be sat 'waiting' to be turned on. This capacity could be in the form of hot water storage contained in an appropriately sized and insulated tank or electrical batteries. With the broadening of the reach of the internet of things, white goods that consume large amounts of hot water or electrical energy could be linked to renewable generation, providing another avenue for the gainful use of energy that might be wasted. The control system therefore needs to have upper and lower indicators lodged (or perhaps learnt by an AI system) into the control system so that these can be responded to. One related issue is that although the control system can be seen as able to respond, these markers for the energy levels and time of day need to be programmed into (learnt by) the system. These are normally associated with assumptions that are in turn based on previous knowledge. Self-learning (or AI) algorithms that now drive some control systems have the capacity to record, and optimise, the control of demand or generation of energy related to building services in buildings. This technology is relatively limited in use at the current time; however, it promises to be the main form of control system for systems that have limited scope for prediction. Controls that are focused on getting the best performance from renewables/STs and buildings are not widely discussed at the current time. However, the smart aspect of controls is the subject of a wider EU piece of research http://www.smartgrids.eu (Smart Grids 2015) that has a broader, gridwide focus. Other published

8.4 Want to know more? Building controls

Wider control strategies for buildings as a whole are described in a variety of texts, the Chartered Institute of Building Service Engineers' Guide H is one that has a wide-ranging and authentically practical view on the use of controls for buildings. This guide explains the more detailed theory behind the operation of proportional, integral, and derivative controls, the use of optimisation in relation to systems starts, and conditions outside the building. There is also very practical guidance relating to building management systems (BMS) and the sometimes difficult issues related to the extension of old systems and integration of old systems with BMS new systems.

Building controls: Realising savings through the use of controls (Carbon Trust 2007) http://www.carbontrust.com/media/7375/ctv032_building_controls.pdf.

Building control systems, CIBSE Guide H (CIBSE 2009).

Camacho, E., Samad, T., Garcia-Sanz, M., & Hiskens, I. Control for renewable energy and smart grids: The impact of control technology (Camacho *et al.* 2011).

guidance is more specific about the integration between the renewables and grids (Camacho *et al.* 2011) and this is supplemented through a series of presentations held at IEEE Control Systems Society Online Lecture Library, http://www.ieeecss-oll.org/. Small-scale renewable energy control systems tend to have a less integrated approach as buildings that use renewables may have one or two generation systems but not the complexity of larger buildings. As the inclusion of renewables/STs become more accepted, systems are likely to develop that cope with the complexity of many-generation systems. Some information relating to the practicalities of these controls on a more individual basis can be found in *Small Scale Renewable Energy Control Systems* (Crowhurst 2007).

8.9   Overheating and controlled ventilation systems

Whilst sometimes it would be reasonable to assume that insulation materials and ventilation systems are mainstream and not necessarily labelled as 'sustainable', unique combinations could be seen as different. The main concept of linked high-insulation levels and ventilation through heat recovery systems is championed by the Passivhaus system (see Chapter 3). In winter conditions when most buildings' windows will not be open, Passivhaus designs conserve heat and supply fresh air taken in via a heat recovery ventilation system. This is one method of reaching internal comfort conditions without the use of major interior heat sources. However, well-sealed buildings and high levels of insulation can be a less satisfactory solution for internal thermal comfort in summer conditions. The NHBC foundation publication, *Overheating in New Homes: A Review of the Evidence* (NHBC Foundation 2012b), states:

> There is increasing evidence that new and refurbished properties are at risk of overheating, especially small dwellings and flats and predominantly single-sided properties where cross ventilation is not possible. However, there is also evidence that prototype houses built to zero carbon standards are suffering from overheating (Reproduced by permission of NHBC Foundation).

Therefore, there is a need for a series of varied solutions to limit overheating in particularly well-sealed new buildings that will still limit winter ventilation heat losses but does not impact the ability to reject heat in the summer months. The section on ventilation in Chapter 3 covers the ability to 'purge' ventilate a building and, if allowed for in such a fashion that does not affect the need to conserve heat in winter months, the overheating issue could be minimised. However, when low levels of thermal mass are inherent in the built form, buildings will have small thermal lag-times which, when combined with high levels of insulation and low ventilation rates, could lead to a more frequent incidence of overheating. In particularly extreme situations, where solar gain is maximised and larger than average quantities of internal gains are experienced, overheating could also occur at times that are normally associated with the need for heating in other more conventional buildings. There is a growing awareness of the issue of overheating in connection with recently built or refurbished highly insulated buildings that are located in climates that do not normally need air conditioning or mechanical ventilation. The UK government through the consultants AECOM has published an investigation of the existing literature of overheating in UK homes (HM Gov 2012). The potential solutions are summarised in Table 8.10, grouped into the levels at which they apply. Note that several blur across the boundaries; for example, are shutters building or equipment?

8.10   Rainwater and greywater collection systems

Rainwater and greywater systems are normally installed in buildings as a supplementary system alongside a 'normal' potable water distribution system within the building. The major issues are therefore to allow both systems to coexist, guaranteeing a safe and continuous water supply whilst making maximum use of non-mains water. Whilst the system may have been designed with information supplied that was current at the tender stages, clients' requirements can change so space and service paths within the building can change.

One of the methods of ensuring that a rainwater system (as opposed to a grey or combined system) operates at an optimum is to focus on the storage capacity of the harvesting system. This, according to BS 8515 Rainwater

Table 8.10 Potential solutions to overheating (HM Government 2012)

Urban	Building	Equipment	Behaviour	Health
Avoid canyons	Cavity wall insulation	Circulation fans	Window opening if external temperatures less than internal temperatures	Drink water, eat cool food
Create blue areas	Chimneys/passive stack ventilation	Curtains	Night ventilation	Sit in the shade
Change building form	External fixed shading	Internal blinds	Reduce bedclothes	Avoid exercise in sun
City albedo	External shutters	Air conditioning of renewable electricity is available at an appropriate time	Reduce clothing	Monitor temperatures for vulnerable people
City ventilation	External wall insulation	Cross ventilation provision	Curtain/blind usage	Follow *Heatwave Plan* intervention levels
Electric vehicles (low noise)	Glazing areas	Grow plants, especially trees	Place vulnerable people with thought—not top floor flats	Obtain ice/cool water supplies
Create green roofs	Internal wall insulation	Ensure mech. vent heat recovery units are correctly operated in summer	Ensure vulnerable people have access to cool/shady areas	Monitor vulnerable people regularly
Tree planting	Low e triple glazing		Turn off lights and non-essential equipment	
Zero energy city	Orientation			
	Solar reflective roof			
	Solar reflective walls			
	Thermal mass			
	Avoid single aspect flats			
	Do not add car parks at expense of green space			
	Consider heating and potential overheating issues in the same package of works			

8.5 Want to know more? Overheating in buildings

Further publications that investigate the issues related to overheating in buildings have been published by the London Climate Change Partnership/Environment Agency, authored by ARUP; another NHBC Foundation publication that is informative when attempting to ensure that homes do not overheat; and a helpful literature review published by the UK Government.

NHBC Foundation. Understanding overheating—Where to start: An introduction for house builders and designers (NF44) (NHBC Foundation 2012a).

NHBC Foundation. Overheating in new homes: A review of the evidence (NF46) (NHBC Foundation 2012b).

London Climate Change Partnership/Environment Agency. Heat Thresholds Project final report, prepared by ARUP (LCCP 2012).

Department for Communities and Local Government. Investigation into overheating in homes: Literature review, AECOM (DCLG 2012).

Harvesting Systems (BSI 2009b, reproduced by permission of BSI), will be a function of the following:

- the amount and intensity of rainfall
- the size and type of the collection surface
- the number and type of intended applications, both present and future

Both building occupants and people working on the water systems in a building need to be able to clearly identify that a pipe or container holds rainwater (rather than drinkable 'wholesome' or potable water). The marking of rainwater supplies is covered by the Water Fittings Regulations (HM Government 1999), which require appropriate labelling, thus preventing contamination of the potable water supply by rainwater. Examples might include a separate low-pressure supply of rainwater intended for a washing machine or an outside tap for garden irrigation labelled as such.

Maintenance of rainwater systems is one of the predictable elements of a yearly cycle. In the autumn, leaves from surrounding trees are likely to collect in the catchment area or in the collection system. This and other debris, biological agents, sediment, and so forth will need to be considered, including inserting cages over vulnerable gullies, sweeping of roofs where health and safety regulations allow for this, and paying attention to filters in various points of a rainwater system. Longer- and shorter-term maintenance will also be necessary. This might be in the appropriate inspection of the water and changing/cleaning of filters/mesh according to the manufacturer's guidance.

Once the system is installed following the relevant building regulations and guidance, including health and safety regulations relating to excavations (see some of the specialist publications at the end of this section), the whole system needs to be inspected and tested:

- Catchment surfaces need to be inspected and cleaned of debris.
- Pre-existing problems related to gutters and down pipes need to be remedied, including supports, clips, and so on.
- External contaminated water must not seep into rainwater harvesting pipework, so appropriate inspection of this aspect will be required.
- Any positions where pipework penetrates the building façade should be adequately sealed so as to minimise an ingress of exterior air.

According to *Rainwater and Grey Water: A Guide for Specifiers* (DEFRA 2007), there are a series of steps that are recommended for the commissioning of a rain- or greywater system:

- Visual check of the pipework systems. Check that actual equipment and pipe layout matches the schematic, or at least there is sufficient similarity between them that maintenance instructions can be followed. Check that all pipes are properly identified and marked.
- Verification of overall system integrity and hydraulic operation using clean water.
- Verification of the operation of control strategy, failsafe features, and indicators using clean water.

There is evidence that failures happen if not properly installed or if the client has not been taken through the operation of any system. According to BSRIA (Brewer *et al.* 2001), when referring to users of rainwater systems, 'They often did not know whether the system was operating or it had failed'. This report was more complementary about installed greywater systems: 'Larger greywater systems were better installed and benefited from professional management and maintenance arrangements'. This report is over 10 years old now, so the technology and systems have matured in the intervening period, but care still needs to be taken when looking after these systems. Modern examples of a direct (Figure 8.5) and indirect (Figure 8.6) domestic scale greywater system are shown below.

Whilst maintenance will ensure the efficient running of any rainwater and greywater system, eventually any of these sustainable water technologies have an endpoint to their life. According to the UK's Chartered Institution of Building Services Engineers (CIBSE 2010), this depends on a range of factors, as can be seen in Table 8.11.

If the information shown in Table 8.11 is shared with those responsible for the maintenance of the system, a procedure with predictive dates and times can be built up to ensure that vital components are either checked at the very least or replaced. (For a more detailed listing of the tasks and the timings of monitoring rainwater and greywater systems, see the guidance listed in Textbox 8.6).

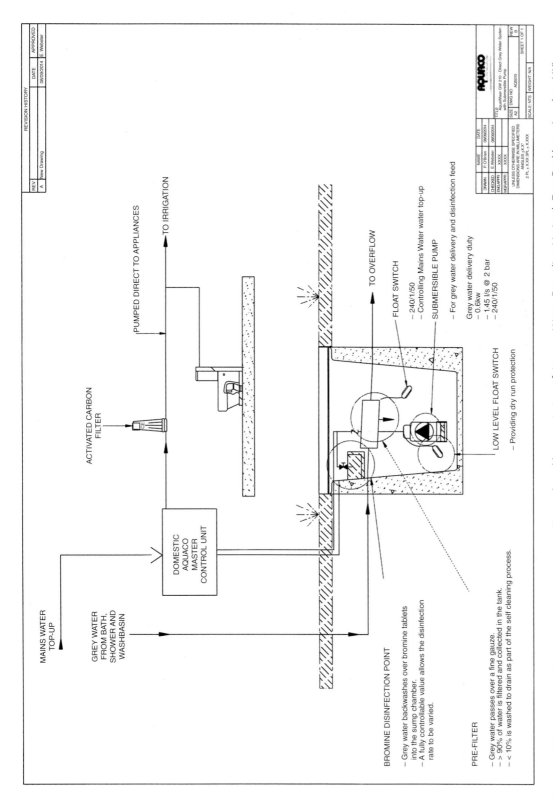

Figure 8.5 Typical direct domestic greywater system (Reproduced by permission of Aquaco Water Recycling Limited, East Peckham, London, UK).

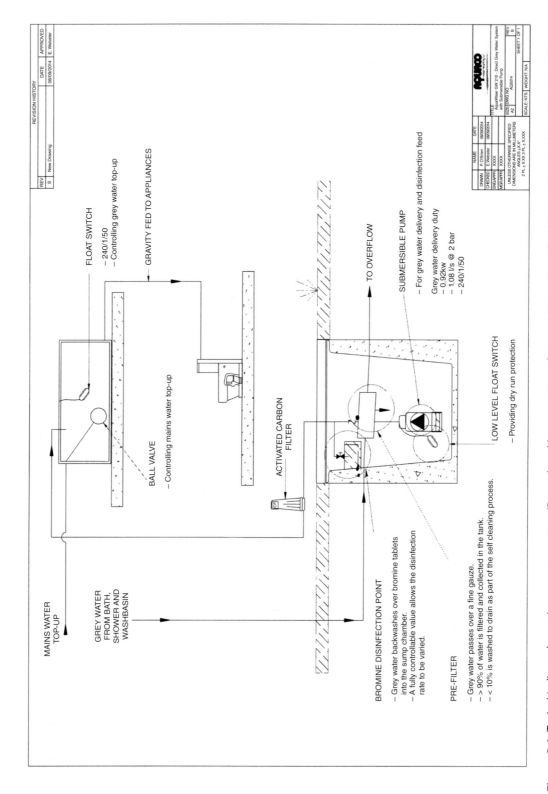

Figure 8.6 Typical indirect domestic greywater system (Reproduced by permission of Aquaco Water Recycling Limited, East Peckham, London, UK).

Table 8.11 The economic life spans of the various components that make up a typical rainwater/greywater installation. (Adapted from CIBSE Guide to ownership and maintenance of building services and quoted in rainwater and greywater: A guide for specifiers, Reginald Brown, BSRIA) table in appendix VI

Element (Years)	Economic Life Factor
Below-ground drainage	35–40
Storage tank (buried or internal)	30–35
Delivery pipework	30–35
Rainwater goods and primary filter	15–20
Delivery pump	8–12
Level sensors	3–10
UV lamp	1

8.6 Want to know more? Rainwater and greywater harvesting

Helpful publications relating to rainwater and greywater systems:

CIRIA C539 Rainwater and grey water use in buildings. Best practice guidance (CIRA 2001).

DIN 1989 Rainwater harvesting systems: Part 1. Planning, installation, operation and maintenance (DIN 1989).

CIRIA C626 Model agreements for sustainable water management systems: Model agreements for rainwater and grey water use systems (CIRA 2004).

Market Transformation Programme. BNWAT05: Rainwater & greywater systems—Supplementary briefing note (MtP 2011).

Environment Agency. Greywater for domestic users: An information guide (Environment Agency 2011).

BSRIA. Technical note TN7/200, Rainwater and Greywater in buildings: Project report and case studies (Brewer *et al.* 2001).

UK Rainwater Management Association website, www.ukrha.org (formally UKHRA).

8.11   Summary

Sustainable technologies are, for the most part, new. Not only to the building designer but also to the specialist trades that are normally assumed to be able to install them. It should not be forgotten that that relative unfamiliarity also resides with the main contractor, the building's owner, and most of all, the occupant. The success of building with sustainable technologies is dependent not only on the theoretical success, carefully calculated at the design stage, but also on the integration of the technology, installation, use, and maintenance of such systems on behalf of all these different viewpoints. This chapter has highlighted some of the more commonly used sustainable technologies currently available. As their use increases the need for such guidance will reduce until someday the more successful technologies will be seen as an obvious fitment on many new buildings.

References

Aberdeen Group (2008) *Building a green supply chain*. Email member.services@ aberdeen.com to obtain a copy.

Blackmore, P. (2010) *Building-Mounted Micro-Wind Turbines on High-Rise and Commercial Buildings (FB 22)*, BRE Press.

Brewer, D., Brown, R., & Stanfield, G. (2001) *Rainwater and Greywater in Buildings: Project Report and Case Studies (Technical Note TN 7/2001)*, BSRIA.

BSI (2004) BS EN 1995-1-2:2004 *Eurocode 5: Design of timber structures*. [Online], Available: shop.bsigroup.com/ProductDetail/?pid=000000000030201375

BSI (2006a) BS EN 12975-1:2006+A1:2010 *Thermal solar systems and components. Solar collectors. General requirements*. [Online], Available: shop.bsigroup.com/ProductDetail/?pid=000000000030216715.

BSI (2006b) BS EN 12976-1:2006 *Thermal solar systems and components. Factory made systems. General requirement*. [Online], Available: shop.bsigroup.com/ProductDetail/?pid=000000000030113770.

BSI (2007) BS EN 15316-4-6:2007 *Heating systems in buildings. Method for calculation of system energy requirements and system efficiencies, Part 4–6: Heat generation systems, photovoltaic systems*. [Online], Available: shop.bsigroup.com/ProductDetail/?pid=000000000030141001.

BSI (2008) BS 7671:2008+A3:2015 *Requirements for electrical installations. IET wiring regulations*. [Online], Available: shop.bsigroup.com/bs7671.

BSI (2009a) BS EN 62446:2009 *Grid connected photovoltaic systems. Minimum requirements for system documentation, commissioning tests and inspection*. [Online], Available: shop.bsigroup.com/ProductDetail/?pid=000000000030141033.

BSI (2009b) BS 8515:2009 *Rainwater Harvesting Systems—Code of Practice*. [Online], Available: shop.bsigroup.com/ProductDetail/?pid=000000000030260364.

BSI (2010a) BS EN 50272-1:2010 *Safety requirements for secondary batteries and battery installations. General safety information*. [Online], Available: shop.bsigroup.com/ProductDetail/?pid=000000000030141654.

BSI (2010b) DD CEN/TS 12977-1:2010 *Thermal solar systems and components. Custom built systems. General requirements for solar water heaters and combi systems*. [Online], Available: shop.bsigroup.com/ProductDetail/?pid=000000000030247841.

BSI (2011) BS EN 50272-1:2010 *Safety requirements for secondary batteries and battery installations. General safety information*. [Online], Available: http://shop.bsigroup.com/ProductDetail/?pid=000000000030141654 [30 Aug 2015].

BSI (2013) BS EN 14825:2013 *Air conditioners, liquid chilling packages and heat pumps, with electrically driven compressors, for space heating and cooling. Testing and rating at part load conditions and calculation of seasonal performance* [Online], Available: http://shop.bsigroup.com/ProductDetail/?pid=000000000030272396 [30 Aug 2015].

Building Control NI (2015) *Northern Ireland Building Control: Regulations*, [Online], Available: http://www.buildingcontrol-ni.com/regulations/technical-booklets.

Camacho, E.F., Samad, T., Garcia-Sanz, M., & Hiskens, I. (2011) Control for renewable energy and smart grids. In: *The Impact of Control Technology*, T. Samad & A.M. Annaswamy (eds.), [Online], Available at www.ieeecss.org [30 Aug 2015].

Carbon Trust (2007) *Building controls: Realising savings through the use of controls*, [Online], Available: http://www.carbontrust.com/media/7375/ctv032_building_controls.pdf.

Carbon Trust (2008) *CTC738, Small-scale wind energy policy insights and practical guidance*, Carbon Trust, London. [Online], Available: http://www.carbontrust.com/media/77248/ctc738_small-scale_wind_energy.pdf [12 Jun 2014].

Carbon Trust (2012) *Biomass boiler system sizing tool user manual* (Version 6.4 3 October 2012), Carbon Trust [Online], Available: http://www.carbontrust.com/media/63116/biomass-software-tool-user-manual.pdf [30 Aug 2015].

Carbon Trust (2015a) [Online], Available: http://www.carbontrust.com/media/147462/j8057_ctl150_how_to_implement_guide_on_ground_source_heat_pumps_aw__interactive.pdf [30 Aug 2015].

Carbon Trust (2015b) [Online], Available: http://www.carbontrust.com/media/147466/j8058_ctl151_how_to_implement_guide_on_air_source_heat_pumps_aw.pdf [30 Aug 2015].

Carpenter, S.R. (1995) When are technologies sustainable? *Society for Philosophy and Technology* **1**(1/2).

CIBSE (2006) TM38 *Renewable energy sources for buildings*, CIBSE, London. [Online], Available: http://www.cibse.org/knowledge/cibse-tm/tm38-renewable-energy-sources-for-buildings [12 Jun 2014].

CIBSE (2009) *Building Control Systems*, CIBSE Guide H, CIBSE, London.

CIBSE (2010) *Rainwater Harvesting: Design & Installation Guide 2010* (Domestic Building Services Panel) CIBSE, London.

CIBSE (2014) *CIBSE Guide M: Maintenance Engineering and Management*, CIBSE, London.

CIRIA (2001) CIRIA C539 *Rainwater and grey water use in buildings. Best practice guidance*, CIRIA, London. www.ciria.co.uk

CIRIA (2004) CIRIA C626 *Model agreements for sustainable water management systems—Model agreements for rainwater and grey water use systems*, CIRIA, London. www.ciria.co.uk

Crowhurst, B. (2007) *Small Scale Renewable Energy Control Systems*, Nordic Folkecenter for Renewable Energy, Denmark. [Online], Available: http://www.folkecenter.net/mediafiles/folkecenter/pdf/Small_Scale_Renewable_Energy_Control_Systems.pdf [12 Jun 2014].

Dangana, Z., Pan W., & Goodhew, S. (2012) Delivering sustainable buildings in retail Construction. In: Smith, S.D (ed.), *Proceedings, 28th Annual ARCOM Conference*, 3-5 September 2012, Edinburgh, UK, Association of Researchers in Construction Management, 1455-1465. [Online], Available: http://www.arcom.ac.uk/-docs/proceedings/ar2012-1455-1465_Dangana_Pan_Goodhew.pdf [4 Apr 2013].

DCLG (2012) *Investigation into Overheating in Homes, Literature Review*, AECOM, Department for Communities and Local Government (DCLG), London, ISBN: 978-1-4098-3592-9, [Online], Available: https://www.gov.uk/government/uploads/system/uploads/attachment_data/file/7604/2185850.pdf [19 Aug 2015].

DECC (2013) *Microgeneration Installation Standard*: MIS 3002 (Requirements for MCS Contractors Undertaking the Supply, Design, Installation, Set to Work Commissioning and Handover of Solar Photovoltaic (Pv) Microgeneration Systems, 3rd ed., updated 01/05/2015 [Online], Available: http://www.microgenerationcertification.org/images/MIS_3002_Issue_3.3_Solar_PV.pdf [30 Aug 2015].

DEFRA (2007) RPWAT03/07 *Rainwater and grey water: A guide for specifiers. Market Transformation Programme (MTP)*, DEFRA.

DIN (1989) DIN 1989 *Rainwater harvesting systems: Part 1. Planning, installation, operation and maintenance.*

ENA (2011) *Distributed Generation Connection Guide (A guide for connecting generation that falls under g83/1-1 stage 1 to the distribution network)* November 2011 Version 3.3. Energy Networks Association, [Online], Available: http://www.energynetworks.org/modx/assets/files/electricity/engineering/distributed%20generation/DGCG%20G83%20S1%20Nov2011.pdf [19 Dec 2013].

EST (2006) *Solar water heating systems: Guidance for professionals, conventional indirect models.* CE131 Energy Saving Trust, London.

EST (2009) *Location, location, location: Domestic small-scale wind field trial report*, Energy Saving Trust, London. [Online], Available: http://www.energysavingtrust. org.uk/Publications2/Generating-energy/Field-trial-reports/Location-location-location-domestic-small-scale-wind-field-trial-report [12 Jun 2014].

EST (2010) *Getting warmer: A field trial of heat pumps*, Energy Saving Trust, London. [Online], Available: http://www.heatpumps.org.uk/PdfFiles/TheEnergySavingTrust-GettingWarmerAFieldTrialOfHeatPumps.pdf [12 Jun 2014].

EST (2011) *Here comes the sun: A field trial of solar water heating systems*, Energy Saving Trust, London.

EST (2013) *Detailed analysis from the second phase of the Energy Saving Trust's heat pump field tria*, Energy Saving Trust, May 2013. [Online], Available: https://www.gov.uk/government/uploads/system/uploads/attachment_data/file/225825/analysis_data_second_phase_est_heat_pump_field_trials.pdf [12 Jun 2014].

English Heritage (2008) *Small-scale solar thermal energy and traditional buildings*, English Heritage, London.

Environment Agency (2011) *Greywater for domestic users: An information guide*, May 2011, [Online], Available: http://www.sswm.info/sites/default/files/reference_attachments/ENVIRONMENT%20AGENCY%202011%20Greywater%20for%20Domestic%20Users.pdf [15 Sep 2015].

ETCUEP (2003) *Technology Transfer: The Seven "C"s for the Successful Transfer and Uptake of Environmentally Sound Technologies.* International Environmental Technology Centre United Nations Environment Programme (ETCUEP), Osaka, Japan. [Online], Available: http://www.unep.or.jp/ietc/techtran/focus/technology_transfer_v6.pdf [12 Jun 2014].

ETSU-R-97 (1996) *The Assessment and Rating of Noise from Wind Farms*, [Online], Available: http://www.hayesmckenzie.co.uk/downloads/ETSU%20Full%20copy%20(Searchable).pdf [12 Jun 2014].

Hammond, G.P., Harajli, H.A., Jones, C.I., & Winnett, A.B. (2012) Whole systems appraisal of a UK Building Integrated Photovoltaic (BIPV) system: Energy, environmental, and economic evaluations, *Energy Policy* **40**, 219–230.

He, M., & Sanders, S. (2011) Design of a 2.5kW low temperature Stirling engine for distributed solar thermal generation. 9th Annual International Energy Conversion Engineering Conference August 8-1, 2011 [Online], Available: http://power.eecs.berkeley.edu/publications/he_design_stirling_engine.pdf [12 Jun 2014].

HM Government (1999) *The Water Supply (Water Fittings) Regulations* [Online], Available: http://www.legislation.gov.uk/uksi/1999/1148/introduction/made [30 Aug 2015].

HM Government (2009) *BERR Wind-Speed Data Base BIS*, HM Gov, London. [Online], Available: http://webarchive.nationalarchives.gov.uk/+/http://www.berr.gov.uk/whatwedo/energy/sources/renewables/explained/wind/windspeed-database/page27326.html [12 Jun 2014].

HM Government (2011a) *Energy Act 2011* HM Gov, London. [Online], Available: https://www.gov.uk/government/uploads/system/uploads/attachment_data/file/48199/3211-energy-act-2011-aide-memoire.pdf [9 Jun 2014].

HM Government (2011b) *Guide to the Installation of PV Systems* (3rd ed. DTI), HM Gov, London. [Online], Available: http://www.microgenerationcertification.org/admin/documents/Guide%20v3.8%20DFC.pdf [12 Jun 2014].

HM Government (2012) *Investigation into Overheating in Homes, Literature Review* Point 47, Executive Summary, AECOM & HM Gov, London. [Online], Available: https://www.gov.uk/government/uploads/system/uploads/attachment_data/file/7604/2185850.pdf [9 Jun 2014].

HM Government (2015) *Building Regulations: Approved Documents*, [Online], Available: http://www.planningportal.gov.uk/buildingregulations/approveddocuments/.

IEC (2004) *IEC 61730:2004 Photovoltaic (PV) module safety qualification*, International Electrotechnical Commission [Online], Available: https://webstore. iec.ch/publication/5739 [30 Aug 2015].

IEC (2005) *IEC 61215:2005 Crystalline silicon terrestrial photovoltaic (PV) modules: Design qualification and type approval*, International Electrotechnical Commission [Online], Available: https://webstore.iec.ch/publication/5739 [30 Aug 2015].

IEC (2008) *IEC 61646:2008 Thin-film terrestrial photovoltaic (PV) modules: Design qualification and type approval*, International Electrotechnical Commission [Online], Available: https://webstore.iec.ch/publication/5739 [30 Aug 2015].

IEC (2012) IEC 61400-11:2012 *Wind turbines: Part 11. Acoustic noise measurement techniques*, International Electrotechnical Commission [Online], Available: https://webstore.iec.ch/publication/5428 [30 Aug 2015].

IPF (2012) *Costing Energy Efficiency Improvements in Existing Commercial Buildings*, IPF Research Programme, [Online], Available: http://www.sweett group.com/wp-content/uploads/2014/03/costing-energy-efficiency-improve ments-in-existing-commercial-buildings-summary-report1.pdf [30 Aug 2015].

LCCP (2012) *Heat Thresholds Project Final Report*, prepared by ARUP, London Climate Change Partnership/Environment Agency, [Online], Available: http:// climatelondon.org.uk/wp-content/uploads/2013/01/LCCP_HeatThresholds_ final-report-PUBLIC.pdf [15 Sept 2015].

Lowson, M.V. (1993) *Assessment and Prediction of Wind Turbine Noise*, ETSU W/ I3/00284/REP.

Lowson, M.V., & Fiddes, S.P. (1994) *Design Prediction Model for Wind Turbine Noise*, ETSU W/13/00317/REP.

Mackay, M. (2012) Wind turbines' lifespan far shorter than believed, study suggests. *The Courier*, 29 December 2012. [Online], Available: http://www.thecourier.co. uk/news/scotland/wind-turbines-lifespan-far-shorter-than-believed-study- suggests-1.62945 [12 Jun 2014].

Milborrow, D. (2010) Breaking down the cost of wind turbine maintenance. *Wind Power Monthly*, 15 June 2010.

MtP (2011) *Rainwater and greywater systems: Supplementary briefing* note BNWAT05 Version 1.0 Market Transformation Programme, [Online], Available: http://efficient-products.ghkint.eu/spm/download/document/id/958.pdf [Sept 15 2015].

NHBC Foundation (2012a) *Understanding overheating: Where to start. An introduction for house builders and designers* (NF44), [Online], Available: http://www. nhbcfoundation.org/Researchpublications/NF44/tabid/515/Default.aspx [31 Aug 2015].

NHBC Foundation (2012b) *Overheating in new homes: A review of the evidence* (NF46), [Online], Available: http://www.nhbcfoundation.org/Publications/ Research-Review/Overheating-in-new-homes-NF46 [31 Aug 2015].

OECD (1995) *OECD Specialist Report: The Global Environment Goods and Services Industry* [Online], Available: http://www.oecd.org/sti/ind/2090577.pdf [13 Mar 2013].

Pennycook, K. (2008) *BSRIA Illustrated Guide to Renewable Technologies*, BSRIA, Bracknell, UK.

Scottish Parliament (2015) *Technical handbooks*, [Online], Available: http://www. gov.scot/Topics/Built-Environment/Building/Building-standards/publications/ pubtech.

SEPEMO (2015) *Welcome to SEPEMO-Build project*, [Online], Available: http:// sepemo.ehpa.org [30 Aug 2015].

Smart Grids (2015) *European Technology Platform for Electricity Networks of the Future, also called ETP Smart Grids*, [Online], Available: http://www.smartgrids. eu [30 Aug 2015].

Thorne, A. (2011) *Solar thermal systems: Key factors for successful installations*, BRE Information Paper IP 11/11 BRE, Watford.

UK Government (2009) *The Town and Country Planning (General Permitted Development) (Domestic Microgeneration) (Scotland)* Amendment Order 2009 No. 34 [Online] Available; http://www.legislation.gov.uk/ssi/2009/34/pdfs/ssi_20090034_en.pdf [7 Oct 2015].

WBC (Woking Borough Council) (2007) *Woking Park Fuel Cell CHP*. [Online], Available: http://www.google.co.uk/url?sa=t&rct=j&q=&esrc=s&frm=1&source=web&cd=1&cad=rja&ved=0CDsQFjAA&url=http%3A%2F%2Fwww.woking.gov.uk%2Fenvironment%2Fclimate%2FGreeninitiatives%2Fsustainablewoking%2Ffuelcell.pdf&ei=_ztJUvmTJNKO7AaT7YDYCw&usg=AFQjCNGgUH75BYAL-oYKj2AyZtjqfCTVJg [12 Jun 2014].

WCED (World Commission on Environment and Development) (1987) *Our Common Future, Report of the World Commission on Environment and Development, Annex to document A/42/427: Development and International Cooperation: Environment*, NGO Committee on Education, United Nations, Geneva, Switzerland. [Online], Available: http://conspect.nl/pdf/Our_Common_Future-Brundtland_Report_1987.pdf [9 Jun 2014].

Zainab, S., Pan, W., Goodhew, S., & Fuertes, A. (2013) Stakeholders perspective on sustainable technology selection to achieve zero carbon retail buildings. In: Smith, S.D., & Ahiaga-Dagbui, D.D. (eds.) *Proceedings 29th Annual ARCOM Conference*, 2–4 September 2013, Reading, UK, Association of Researchers in Construction Management, 1219–1229.

9. Future of sustainable construction

9.1 Future directions and policies

Predictions are a challenging area, as has been identified by a number of researchers, including the Danish physicist Neils Bohr, writers such as Mark Twain, and sportsman Yoggi Bera: 'It is dangerous to make forecasts, especially about the future' (Wolf *et al.* 2000). Accurately forecasting the future, especially aspects of such a diverse field as sustainable construction, is fraught with difficulties. These difficulties often lie in the areas of changing technology, climate, politics, consumer behaviour, and the way we live our lives. Henri Poincaré, however, stated about the use of hypotheses, 'It is far better to foresee even without certainty than not to foresee at all' (1913). And so, taking Poincaré's lead, this chapter analyses a number of areas that will influence the future of sustainable construction.

9.1.1 Climate

One relevant issue for the future of sustainable buildings is the strong likelihood of increasing extremes in climate, defined by NASA (2005) as a term to describe 'how the atmosphere "behaves" over relatively long periods of time'. Whilst weather, which refers to the short-term conditions of the atmosphere (NASA 2005), may indeed be too changeable to predict very far ahead, climate, because it relates to a longer time span, can be predicted to a greater degree, and reactions to relevant and extreme elements need to be factored into our buildings. This is easier said than done and, where our buildings have long lives, they are very likely to be subjected to stresses imposed by a changing climate. The UK Met Office predicts that, within the lifetime of buildings being constructed now, precipitation will increase up to 10% in the northern UK, and some southern areas could experience decreases of up to 5%. The UK is projected to experience temperature increases of up to around 3°C in the south and

Sustainable Construction Processes: A Resource Text, First Edition. Steve Goodhew.
© 2016 John Wiley & Sons, Ltd. Published 2016 by John Wiley & Sons, Ltd.

2.5°C further north (Met Office 2011). These predictions, combined with the increased likelihood of extreme winds, will challenge the performance of our buildings if they are built to current standards and according to today's acceptable practices. The future will require adaptation, namely, adapting many buildings to ensure that they function as well 50–60 years from now as they do today. Bill Gething and Katie Puckett, in their book *Design for Climate Change* (2013), predict that there may be some legal basis for litigation concerning the design of buildings where the probability of future climate change is a known factor.

9.1.1.1 Comfort, construction, and water

In 2010 the UK's Technology Strategy Board, now known as Innovate UK, launched a programme that aimed to explore how buildings need to be designed and constructed to respond to the most likely changes in the UK's climate. This programme 'Design for Future Climate' (Innovate UK 2015) enabled 50 design teams to consider the impacts of these changes in climate, such as hotter drier summers, warmer wetter winters, more extreme weather events and a rise in sea level. When translated into building-related actions the findings were categorised into three main topic areas: comfort, construction and water, as shown in Table 9.1.

Within the 50 case studies investigated in the Technology Strategy Board's programme there is a broad assortment of project types, locations, and clients, making for a rich set of data which can be viewed online (Innovate UK 2015).

Another take on future climate is the Chartered Institution of Building Services Engineers' (CIBSE) publication *Sustainability, CIBSE Guide L* (Cheshire & Grant 2007), which describes some of the extra measures that could be taken so that UK buildings can adapt to the most prominent climatic possibilities, shown in Table 9.2.

As can be seen from Table 9.2, a wide range of advised actions relating to buildings and their immediate surrounds flow from the predicted climate change effects. Differences in ranges of outdoor temperatures lead to changes in the environments of indoor spaces, testing the resilience of construction materials and details for thermal expansion and

Table 9.1 Climate and building-related actions

Future Climate-Related Design Category	Major Issue According to the Technology Strategy Board's Design for Future Climate Programme
Comfort	Interior and exterior overheating, with some lesser more localised issues of underheating
Construction	Stability above and below ground, fixing and weatherproofing of buildings, the behaviour of materials in extreme conditions and the working conditions on construction sites
Water	Water conservation, drainage, and the interlinked issue of flooding, landscape

Table 9.2 Key predicted climate changes for medium-high emissions scenarios (from Cheshire & Grant 2007, reproduced by permission of CIBSE)

Key Predicted Climate Change Effect	Description	Implications for Buildings	Example Measures
General increase in temperatures	Current modal temperature in London is 10°C; predicted to be 12°C by 2080 High summer temperatures will become frequent, and very cold winters will become rare.	Increased outgassing of pollutants from structure and furnishings affecting indoor air quality Decaying waste more likely to smell and may cause problems with infestation	Avoid high internal temperatures and VOCs in finishes, construction materials, carpets, and furnishings; select labelled low-emission products. Allocate adequate space and ensure secure, sealed storage of segregated wastes in regularly cleaned area designed to avoid overheating.
Milder winters	Mean winter temperatures predicted to increase by 2.5°C in SE England and 2°C in N England. Decrease in heating degree-days (cf 1980s) of 35%–40% for London, 30% for Edinburgh by 2080.	Reduced energy use in winter	Consideration should be given to the potential to reduce system capacity and thereby reduce energy use during winter months.
Rising summer temperatures	Increase in mean summer temperature: (a) SE England: 2.5°C–3°C by 2050; 4.5°C–5°C by 2080 (b) N England and Scotland: 1.5°C–2°C by 2050; 2.5°C–3.5°C by 2080 Occurrence of temperatures >28°C increases from 1–2 days/year (1989) to 20 days/year by 2080 in London. Increased occurrence of hot spells (no. of days when previous 3–5 days have >3 hours at 25°C). Increase in cooling degree-days (cf 1989); e.g., +150–200 for London, +20–25 for Edinburgh, by 2080.	Higher summertime temperatures will increase overheating risk in buildings. Due to rising external temperatures, the traditional mechanism for cooling buildings through ventilation with external air cannot be relied upon. Careful design required to reduce dependence on mechanical cooling and to maintain indoor comfort.	Incorporation of intelligent ventilation systems, preferably automated, are set up to limit ventilation during warmer parts of the day and recharge 'coolth' reservoirs in high-mass buildings when external air temperatures permit.

(Continued)

Table 9.2 (Continued)

Key Predicted Climate Change Effect	Description	Implications for Buildings	Example Measures
Enhanced urban heat island effect	In central London the urban heat island effect can currently lead to elevated summer nighttime temperatures 5°C–6°C warmer than temperatures in rural areas outside London. This effect is expected to intensify due to climate change, leading to a greater temperature difference between the heat island and surrounding rural areas, and more hours of nighttime heat island effect per year.	The increase in nighttime urban temperature due to the urban heat island effect reduces the ability of buildings within the heat island to use nighttime cooling as a strategy. Increased temperatures, particularly at night for any building within urban conurbations, reduces the ability to dissipate heat at night, making nighttime 'free cooling' less practicable in the future.	Planting of trees, vegetation, and green space for shade and natural cooling through evapotranspiration (estimated to result in 1°C–5°C reduction in peak summer temperatures) Green roofs (may reduce surface temperature by 20°C–40°C compared to a conventional dark flat roof)
Wetter winters, more intense rainfall	Winters will become wetter by up to 10% across the country by the 2050s and up to 20%–30% across the UK by the 2080s. In combination with higher wind speeds, occurrence of driving rain will increase in winter months.	Inability of drainage system to cope with storm surges, damage to some building materials. Increased risk of flash flooding.	Undertake flood risk assessment; design for flood resilience and to reduce flood risk. Use SUDS* techniques. Locate 'resistant' uses (e.g., car parking) in high-risk areas and raise ground floors. Use of building materials that are resilient to driving rain
Drier summers	Decrease in summer rainfall of 40% across most of the UK by the 2080s. Decrease in soil moisture content, especially in summer; 30% decrease predicted for most of England and 10%–20% for the rest of England, Wales, and most of Scotland	Increased pressure on water resources and increased occurrence of hosepipe bans. Increased pressure for water storage capacity. Possible disruption due to increased subsidence due to drier clay soils.	Apply water efficiency principles in new and existing buildings. Plan for building operation in drought conditions. Avoid siting 'water-hungry' developments in areas prone to drought.
Higher daily mean winter wind speeds	Possible increase of 7% by the 2080s, with the greatest increase in SE England	Increased risk of wind damage. Higher infiltration and winter heat loss. Pylons carrying electricity and telecommunications may be vulnerable to higher wind speeds. Increased level of power shortages and outages.	Strengthening of tall buildings, increased airtightness, and incorporation of cladding materials able to cope with higher wind speeds Incorporation of local/on-site generation of power and renewables to reduce dependence on the grid

UV degradation. Greater rainfall and the predicted incidence of higher wind speeds will combine to test the appropriateness of existing building and landscaping designs.

These drivers have led to the publication of new draft international standards, including the first draft of a new standard related to sustainable communities. ISO 37101, entitled 'Sustainable Development and Resilience of Communities—Management Systems—General Principles and Requirements', outlines the requirements, provides guidance, and points the reader towards a number of potential supporting techniques and tools to ensure a good level of sustainable development in communities (ISO 2014b). The document is designed to help all kinds of communities manage their sustainability and resilience as well as improve the contribution of communities to sustainable development and assess their performance in this area. At the time of writing, the committee draft version of ISO 37101 could be purchased, allowing a preview of its contents before the final publication date, which is set for 2016 (ISO 2014a).

9.1.2   Future homes

Climate and the assumption that it will have a greater influence in the future is just one area that impacts building adaptation. The ability to meet anticipated performance levels in future buildings is another influence on the construction sector. A National House Building Council (NHBC) Foundation report focuses on the production of future homes and concentrates on the factors that will enable these homes to deliver their full potential. The main issues are quite familiar: policy, planning and context, building fabric, integrating services and mechanical ventilation, and heating and low carbon technologies. However, the report does diverge from some current guidance in that it advocates a more root and branch change in the design of housing. The report describes previous attempts to evolve the design of buildings as one of the reasons for future homes not meeting zero-carbon goals. Taking more or less pre-existing home design and adjusting the fabric to accommodate more thermal insulation or more sustainable technologies is almost certainly not the best basis for an optimum solution. The report advocates a fabric first approach, one that is familiar from the major concepts described in Chapter 3 of this volume. It advocates ensuring that new homes have enough space for services, a large southerly roof area, little overshadowing, adequate ventilation, realistic wall thicknesses, and optimal window areas for daylighting, without causing overheating (NHBC 2013).

9.1.3   Smart homes

Producing comfortable, low-carbon homes fulfils one criterion towards a possible future for sustainable construction, but the rapid increase in the capacity and capability of interactive communications offers a different angle

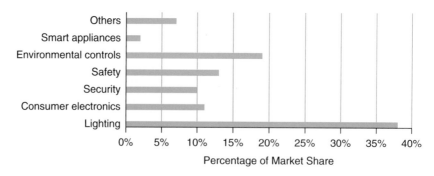

Figure 9.1 Market share for appliances and smart home applications (BSRIA 2013, reproduced by permission of BSRIA).

on that future. The Building Services Research and Information Association (BSRIA) believes that the market for appliances and home infrastructure will reach €620 million in Europe by 2015, growing at a rate of 8% per year (2013) (Figure 9.1). The connected and smart home product market grew by almost 19% in the period 2010–2012 to reach just over €510 million, and it is estimated to grow by an average of 8% each year until 2015.

It is envisaged that the current connected and smart home market will broaden from a niche high-end market to one that sees growth in semi-domestic commercial applications. It is forecast that the use of smart controls for lighting will be the main area of growth for the domestic market, with environmental controls seeing the second highest amount of growth. BSRIA noted that the assisted living home is an important market in the Netherlands and obtains financial support from the Dutch government. In France the current thermal regulation, RT2012, imposes the monitoring of energy use in dwellings, and therefore, the market for home energy management systems (HEMS) is expected to grow between 2014 and 2016. There are many new HEMS products that are likely to feature as standard equipment in future sustainable homes. One of the issues for these products is ensuring that they are able to operate satisfactorily with energy and heating systems produced by a wide range of manufacturers. For the smaller suppliers of HEMS this tends to require partnerships with energy suppliers to provide HEMS products. An example of such products includes a group of intelligent residential environmental controls, including intelligent thermostats, automatic radiator valves, and electric heating control systems. Whilst this array of different technologies within homes may well increase the number of homes identified as 'smart', the tightening of energy efficiency legislation for new buildings in many markets, particularly in France, will probably increase the penetration of such technologies.

9.1.4 Smart cities

According to the United Nations (UN 2012), in the time between 2011 and 2050, the world population is expected to increase by 2.3 billion, passing from 7.0 billion to 9.3 billion. At the same time, the population living in urban

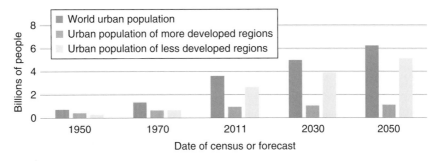

Figure 9.2 Global urban populations by development group, 1950–2050 (UN 2012).

areas is projected to increase by 2.6 billion, passing from 3.6 billion in 2011 to 6.3 billion by 2050 (see Figure 9.2). These predictions point towards population growth becoming an urban phenomenon concentrated in the developing world (Satterthwaite 2007). The future implications of this expansion in urban areas have repercussions on the way sustainable construction will develop between now and 2050. In 2011, 23 urban agglomerations qualified as megacities because they had at least 10 million inhabitants, a global total of 359.4 million people (UN 2012). This number of megacity inhabitants is predicted to nearly double by 2025 and reach 630 million, resulting in one person out of every seven or eight living in an urban area residing in a megacity. Whilst Tokyo qualifies as the largest current megacity with a population of 37.2 million people (2011 figures), it has grown to this size over a number of years, and the growth is partly due to the urban agglomeration of 87 surrounding cities and towns, including Yokohama, Kawasaki, and Chiba (UN 2012). Will future megacities follow the same pattern, growing 'through acquisition' rather than wholly new construction? The implications of the differences between these two patterns can be significant; the construction of infrastructure from scratch or amalgamation with pre-existing patterns of growth can both lead to sustainable solutions, but each needs a different strategy. The sustainability of the largest cities in the world—megacities—and the buildings that make up those urban areas may well be judged differently to buildings in smaller conurbations.

As Tokyo combines the largest population on earth with some of the most advanced technology, this is a good case study to use in order to predict the performance of future megacities. The pressure on space illustrated by the cost of land in central areas of Tokyo is very likely to result in vertical cities with even higher densities than are currently experienced. Just as smart homes are a likely product of the logical extension of current technological innovations, dense vertical cities of smart, sustainable buildings are the likely future of cities. Buildings will need to help provide food and energy alongside the more traditional internal comforts, security and a place to live and work. For buildings to carry out these extra duties, they will need to react to changes in their surrounding environments, both external and internal.

According to a report by Arup (Buscher et al. 2010) the building of the future will be as follows (descriptions reproduced by permission of Arup):

- *Flexible*. Buildings will be prefabricated and modularised to enable continuous flexibility and adaptability so that components can be installed, repaired, and upgraded by mechanised and automated operations. They will have a semi-permanent structure that can be adapted over the life of the building to enable adaption to cope with a change of use for example. Occupant specific components such as the façade, walls, finishes and mechanical plant will be designed to be replaced or uprated after a decade or so. Loose fit-out elements, such as IT infrastructure, will need to be able to accommodate rapid change.
- *Sustainable*. Future city buildings should be capable of producing more energy, water, and food than the occupants consume. Vertical farms and urban agricultural systems, traditional and hydroponic, alongside and including green spaces for the city's inhabitants, can be integrated into these vertical communities. Buildings will generate energy through a range of solar and other sustainable technologies, such as creating biofuel from algae grown by façade-based solutions. This form of adaption of buildings and the ability for buildings to react to changing external or internal conditions normally implies high-tech solutions, but the use of microalgae to improve the levels of shading through the use of greater concentrations of algal growth is one part of the experimental building façade shown at the IBA show in Hamburg in 2013 (IBA 2013). Other developments include water drawn from humid air using modified wind turbines, and waste water might be recycled using the food-growing systems.
- *Smart and integrated*. Buildings will be linked to the surrounding infrastructure to provide an integrated network of transport and utilities. Optimisation of resources will be achieved by collecting data on an individual building's energy consumption and combining it with data from other buildings. A web of green spaces could be formed by linking buildings via green pedestrian bridges.
- Smarter buildings represent a logical direction for the future; however, the ugly truth is that many people do not have a home to call their own. Future policies relating to empty homes are likely to encourage or perhaps even cajole owners into making them available. These policies may also impact empty spaces above shops and commercial-type premises that could feasibly be used as homes. The UK government has allocated funding to tackle long-term empty properties that would not come back into use without additional financial intervention (DCLG 2012). It was hoped that this initiative would provide 3,300 affordable homes by March 2015, and if successful, the intention was that it would result in a reduction in empty properties. No new information that the author can locate is currently available at the time of writing to firmly establish whether this goal was met. A more ambitious initiative run by the UK's Empty Homes Nationwide (EHN 2013) aims to bring 100,000 homes back into use over the next 15 years. Phase one involves a pilot project to refurbish 300 vacant properties by the end of April 2014. At the time of writing, the results from this project are still unknown. According to the UK Government's Homes and Communities Agency, empty homes in England account for three percent of the total housing stock, which translates into a little fewer than 750 thousand

dwellings (DCLG 2012). It is likely that 'hard to treat' empty properties will be targeted by future initiatives as the 'easier' properties are occupied.

The future of commercial buildings may follow the lead of Marks and Spencer's Plan A and their flagship store built in Chester, England. Another retailer, Sainsbury, has also opened what they call a 'triple zero' store in Leicester. The fundamental underpinnings of this store are zero carbon emissions from operational energy, zero waste to landfill, and zero impact on water use within the local catchment area. Neil Sachdev, Sainsbury's property director, has been quoted proposing, 'We aim to be the UK's greenest grocer and achieve our 20 × 20 target to reduce our operational carbon emissions by 30% absolute' (Sainsbury 2014).

9.2   Future technologies

Holistic changes to buildings and cities will have major impacts on the sustainability of the built environment, and these are likely to be made up of many smaller changes in policy and technologies. There have been a number of reports that have adopted existing technologies from spheres other than the design and construction of buildings and predicted possible future technologies for homes and workplaces. A joint project, Imagine 2050, between the London School of Economics and Veolia, an environmental services company, focuses on these future technologies (Veolia 2013). The report in which the project is presented postulates two possible scenarios: one based on more communal living and exchange and the other based on self-reliance and a less-collective society. The end result is a prediction that system planning and the use of new technologies would allow city emissions to be reduced by 80% (since 1990), compared to 40% in an alternative resource-hungry future scenario. Water consumption would be a quarter less (100 litres per person per day instead of 130 litres). The forms of technology that the report predicts for the 2050 resource-efficient home includes a kitchen where waste is sorted by nanoscopic robots and food packaging that is designed to degrade in line with sell-by dates. 3D printers would play a prominent role in producing custom-made objects for the home alongside new paints and materials that optimise natural light and improve energy conservation. Whilst many of these features would, indeed, contribute to more sustainable homes, the occupant behaviour that drives the use of such technologies might counter some of the predicted gains.

As the number of appliances and devices that have a capability to communicate and form part of a network increase year by year, it becomes possible for the technologies across a whole building to be linked. By 2020, it is estimated that there will be 50 billion networked appliances and sensors across the world, forming the 'internet of things'—a global network of data-generating devices, sensors, and their associated URLs (RAE 2013). Within buildings that have building management systems, the communication between sensors would enable the system to respond to individual occupant preferences for environmental variables such as light and temperature settings. The Royal Academy of Engineering (RAE 2013) also highlights the possibility of using networks of devices to care for the elderly or the infirm in their own homes rather than assuming that these vulnerable people need to be in hospital. However, as the devices around a building become able to communicate and react, so the building's occupants become reliant on the operations of these technologies. If these technologies fail either mechanically, electrically, or as a result of communication failures, the function of the building is impaired. As Professor Doug King, chief science and engineering advisor at the UK Building Research Establishment (BRE) stated when referring to the failure of an automatic shading device at the BRE: 'What was commissioned as a smart building became dumb' (RAE 2013).

Smart home energy management is also likely to become more widespread as a technology in the future. It offers the energy supplier the opportunity to switch power usage to those devices that can use electricity at times when other high-use operations are not functioning. This can and has been achieved using a simple tariff system, but future technological spread may enable a wide range of electrical uses to be switched on automatically by a power company. This may also allow favourable renewable sources of energy (a good solar day or night of wind generation) to be capitalised upon, allowing tariffs to be adjusted accordingly, thus lowering the demand for all-important peak use periods.

One unintended issue that may result from the heavy use of existing wireless communication systems could be the eventual run-out of capacity using existing technologies. In the future LI FI could offer low energy and low cost data transfer as well as allowing for light and information to be transmitted between homes and people. LI FI is a wireless form of communication that uses visible and near visible (UV and IR) light instead of radio waves. LI FI light-emitting diodes (LEDs) can communicate considerably more data than radio waves, and their operation is also more efficient. Additionally, LEDS are more secure and convenient than radio waves as they too use part of the electromagnetic spectrum: visible light. LI FI uses visible light from LED bulbs with an amplitude that is modulated at very high speeds with changes that are not visible to the human eye (TEDGlobal 2011). In this way information and visible light can be transmitted simultaneously using micron-sized LEDs. These LEDs are much smaller than the commonly used 1,000 micron-sized LEDs and can fit into the space occupied by a single 'normal' 1 mm^2 LED. Thus, a 1 mm^2 array of tiny LEDs could communicate one million times as much information as a regular 1 mm^2 LED.

LEDs will also continue to play an important part in the future illumination of buildings. A number of the largest lighting equipment manufacturers are producing LED tube replacements that can save up to 60% of energy compared to the most efficient fluorescent tubes. These are coming onto the market at the time of writing, and if these replace fluorescent tubes in the majority of locations, namely offices and industrial buildings, it could have a considerable impact on energy consumption directly connected with lighting. A further gain is also possible in terms of a reduction of heat load on air-conditioning systems, due to the fact that LEDs run cooler than fluorescent tubes.

With the need for tomorrow's megacities to accommodate vast numbers of people, the requirement for bigger and indeed higher buildings is becoming ever more necessary. Lifts (elevators) that can travel to the top of the tallest of buildings are significant in terms of building services. A simple but very effective technological upgrade is to improve the strength and reduce the weight of elevator cables, enabling more storeys to be connected and reducing the energy use of the elevator system.

Future technology is likely to be innovative in the way it is used (rather than changing fundamentally). District heating, larger combined heat and power plants, commercial-sized wind turbines, and larger solar installations are already being shared by communities in many parts of the globe. A recent report from the think tank ResPublica predicts that, through interventions, the amount of power generation resulting from community ownership could rise in the UK from the existing 60 MW in 2013 to 5.75 GW by 2020 (Harnmeijer *et al.* 2013). The sorts of incentives or interventions being envisaged include financial backing by local authorities and the extension of generation tariffs. These could be accompanied by community energy systems being part of energy development plans alongside the establishment of community commissions to act as ways of encouraging communities to take part in solving local energy questions. This element of community involvement may also play a part in the UK Government's Department of Energy and Climate Change (DECC) programme that is designed to develop new heating and cooling networks and expand existing networks (UK Government 2014). This may have some direct impact on future sustainable construction as the programme encourages schemes aimed at drawing their heat energy from renewable, sustainable, or recoverable sources. This includes quite a broad range of heat sources, such as waste heat from industry, energy from waste plants, and biomass combined heat and power. The programme's inclusion of commercial backing may well open up this form of heat network for a much wider range of buildings than is currently included and spread the skills and knowledge necessary to link communities with their energy generation and distribution.

One such heat network, based in Kingston, London, uses water from the River Thames as a source for heating 137 apartments (United House 2013). The project heating system uses a water source heat pump system to provide both space and water heating. A pump house draws an average of 6.5 m litres of water per day from the River Thames, at an average temperature of 9°C, and processes it through a heat exchanger. It is predicted that the system will save almost 500 tonnes of CO_2 and reduce household heating bills by 20%–25% over conventional gas boilers.

Sources of heat can be viewed as relatively plentiful; however, storing electricity is more difficult. Nissan has explored and carried out a successful early field test on a system that will allow buildings to use the power from the electric batteries of Nissan plug-in electric cars. Nissan has termed this innovation a 'vehicle-to-building' (VTB) system and currently allows up to six electric vehicles to be connected to a building's power distribution board to help regulate electricity bills (Webb 2013). The VTB system charges the electric vehicles when the power from the grid is not at peak demand and uses the cars as a storage medium from which it can draw power when the grid is at peak capacity. This is cost effective, but more importantly, it is an organised method of getting larger, useful quantities of electrical power, which would be impossible with smaller numbers of vehicles. The impact upon the grid is also beneficial; if this system were more prevalent, it would minimise the need for generation capacity at peak times. According to Nissan, the VTB system has been used at the Nissan Advanced Technology Centre in Atsugi City, Japan, since July 2013, leading to a 2.5% reduction in electrical power use during peak hours. Whilst this might seem modest, the impact an electrical generating grid could have on efficiency may well be larger, especially if more cars could be used as part of the system.

Whilst metered supplies of potable water are commonplace, future measurement of waste water might be just as familiar. An innovate UK project undertaken with the UK's Wessex Water utility company has been trialling a metering technology that can assess the amount of waste water discharged by commercial premises. The occupants can then be charged the amount for the sewage they produce, and the hope is that this form of technology will lead to commercial companies reducing the amount of discharge from their premises (by using other innovative technologies and processes). This might include greater use of water recycling, rainwater storage, and soakaways (WWT 2014). Technologies such as this can also play their part in reducing our water footprint, the subject of a new international standard. ISO 14046:2014 *Environmental Management—Water Footprint—Principles, Requirements and Guidelines* will enable interested parties to measure and report their water footprint as a stand-alone study or as part of a wider environmental assessment (ISO 2014c). This may promote a wider interplay between water use and practices involved in construction, since footprinting measurements reveal a fuller picture about water usage associated with construction practices, choice of materials, and technologies.

9.3   Future energy

The future of energy within the sphere of sustainable construction is a very broad area. Many different types of energy impact buildings, for example, electrical energy, heat, and mechanical and chemical energy, and the impact of their use can be assessed in terms of a single house or a whole city. Some commentators are predicting the development of stand-alone power solutions, with each building contributing to its own and also communal needs. Others advise that heat networks driven by a central source are the future. In truth, a mixed economy is most likely. Whatever solutions are implemented, newly built or refurbished buildings will have to adjust their design, construction, and care in line with future energy supply and generation.

Focusing on more immediate futures, changes in the supply and therefore the price of energy will change the behaviour of designers and the occupants of buildings.

The way in which the construction industry procures, designs, constructs, and manages buildings can change due to predicted external pressures that are economic in nature, as well as other changes. One of the major influences on the economic activity of most sectors within most countries is the price of energy. An article in the magazine of the UK Royal Institute of Chartered Surveyors, *Modus* (RICS 2013), focuses on the predicted outcomes of an extreme change in the price of crude oil. Predicting the future direction of sustainable construction within the constraints of this scenario is not exact, but it is possible to estimate some patterns of activity change. Certain events, such as a dramatic increase in the price of a commodity such as crude oil that is felt to be vital for construction and use in buildings, may well change patterns of behaviour. Whilst a range of different futures is possible, it may be prudent to consider some of the potentially impactful events that are likely to occur given such a scenario and consider the continuing impact upon what we do as far as construction is concerned.

The possibility of a collision between a tanker and another vessel or similar event blocking a major transport route would have a considerable impact. The RICS piece imagines such a future and quantifies the possible impact upon a world coping with oil at US$200 for every barrel (even if this feels a little remote at the time of writing). There are very predictable impacts on most parts of the economy such as transport or agriculture. There are also potential impacts on the costs of construction including the cost of materials and, in particular, items such as steel, concrete, plastics, and asphalt. The impact on land values is also discussed and, according to Ian Bailey, there are correlations between land prices and the price of crude oil. The competition between the ability to use land for farming and for property affect the choice of building location. John Lovell, the global real estate sustainability leader at Deloitte, predicts that global retailer clients could spend £400 million more on energy over a 10-year period, taking into account future price rises and carbon taxes. Energy performance will then become more central to the value of real estate and inefficient properties; those that use considerably more energy per square metre are likely to suffer an increased form of depreciation. Mr. Lovell approximates that we waste 89 out of every hundred units of energy, which means that efficiencies should lead to significant savings. In combination with these efficiencies, one of the positive aspects of US$200 a barrel for crude oil would be that global carbon emissions would be extremely likely to drop, allowing us to avoid some of the most dire predictions relating to climate change resulting from emissions.

As with most economic drivers, scarcity can drive the search for alternative supplies to meet demand. Peak oil is a term referring to the point in time when the maximum rate of extraction of crude oil is reached, levelling off soon afterwards and then declining. This was expected to occur around 2010, and this anticipated scarcity will probably drive the development of new technologies such as fracking, which will enable other reservoirs of fossil fuel energy to be tapped. According to the British

Petroleum (BP) statistical review of world energy 2013, Brent crude went from $12.8 a barrel to $28.5 a barrel between 1976 and the year 2000. However, a considerable increase took place from 2005, when Brent crude was worth $38.27 per barrel, and increased in three years to $97.26 per barrel; the price then fell back to $61.67 per barrel in 2010 before increasing and hitting a high of $111 a barrel, which has been relatively static since 2011. Whilst some prices, particularly those for petroleum spirit sold to motorists in the UK, are heavily taxed, a doubling of the price from $111.26 per barrel to over $200 a barrel would have a considerable impact on major parts of all economies. As if the unpredictability of predictions needs any emphasis, the relatively unexpected reduction in the price of Brent crude to near $80 a barrel in late 2014 underpins the notion that we cannot predict energy prices with accuracy so resilience and flexibility in our energy use should be one of our future priorities.

9.3.2 Future sources of heat energy for buildings

As cities become denser, heating networks and associated power generation are likely to become normal practice. Larger quantities of heat can be obtained from water sources in the future; these sources have advantages over the use of ground and air sources that can vary in temperature more readily with changes in the external environment. One relatively constant source of heat in larger cities is the large amount of warm air in underground train tunnel networks. This source of heat is to be exploited by Islington Council's Bunhill Heat and Power heat network, which already supplies 700 homes (London Mayor 2013). This project will utilise local sources of waste heat from a London Underground ventilation shaft that is connected to the Northern Line. A number of other opportunities to direct heat into existing or new heat networks have been identified in a report by Buro Happold for the Greater London Authority in early 2013. The report selected a number of emerging heat source areas according to the following criteria: the availability of multiple secondary heat sources, the heat demand density suited to such sources, and a location close to existing or planned heat networks. The emerging areas that have been selected are shown in Table 9.3.

A variation on this theme sees Glasgow planning to use water from abandoned mines as a heat source providing under-street heating similar to that used in Hamburg and Stockholm. It is estimated, through data from

Table 9.3 Areas of London aligned to probable future sources of energy

Location	Rationale
Brent Park	Data centre and transformer stations supply
Paddington and Farringdon	Demand well suited to low temperature sources
Edmonton	Low carbon power station supply
Barking and Royal Docks	Multiple sources, existing network forecast, extensive new build
Hounslow	Potential for high supply from environmental sources

the British Geological Survey (BGS 2013), that the water could provide the heat (using heat pump technology) to service 40% of the city's heating requirements.

Heat can also be recovered from wastewater in sewers. Wastewater—a combination of water that gets discharged, a mix of hot water from toilets, sinks, showers, dishwashers, washing machines, and so on—has a relatively constant temperature, higher than that of water taken into buildings. SHARC Energy Systems uses a heat pump to transport the heat from that wastewater and transfer it to the flow of potable water that enters domestic and commercial properties. Through the use of heat exchangers and a number of different looped systems, the foul effluent is separated from the clean potable water flow (SHARC 2014).

9.4   Future materials

Whilst future homes and cities can play a large part in influencing future construction, sometimes an apparently smaller event, like the introduction of new materials, can be just as influential. Just as the cheaper efficient transportation offered by railways has brought new mass-produced construction materials to the more remote parts of the UK in the late nineteenth century and altered the scale, scope, and design of homes and commercial buildings, so innovative materials of the future can change the landscape.

One aspect of future buildings and construction hinges upon the fact that the building stock has a very low turnover, and the need to improve existing properties is more important in numerical terms than innovative new builds. Existing buildings already have a defined footprint, and normally most improvements need to take place within that footprint. Higher insulation values that are being demanded from retrofitted properties, therefore, create a problem. How should one insulate but still maximise the defined internal areas of a property, when external insulation is not an option? Most of the major manufacturers of insulation are looking to improve their products, not only by lowering thermal conductivity values but also by the subsequent reduction of the thickness of insulation products to overcome this issue. Many manufacturers are developing new insulating materials that promise to be thin, lightweight, and available in panels, and there are numerous questions about these new materials. What will such a material be like? How will it perform? Will the production process be environmentally benign? Will the hygrothermal requirements of a building be taken into account in the material's installation and in situ performance?

At the point of writing, many of these products are not yet ready for market, but some will feature organic aerogel in panels and quilts and promise to be 25%–50% slimmer than conventional alternatives. Future performance will be judged against the claimed thermal conductivity value of below 0.16 W/mK, which is lower than the generally accepted lowest claimed thermal conductivities for existing high-performance thermal insulation products at 0.2 mW/mK. This area of development can also easily be modified to suit different sectors with the highest performing products demanding the highest prices. These materials may provide building designers with the ability to renovate buildings that are currently thought not to be economically feasible projects due to size constraints and issues in meeting thermal regulations.

Other thin insulation products that may replace existing traditional materials could lead to a marked improvement in thermal values in some of the most difficult markets, such as heritage buildings. The properties and importance of these sorts of materials need to be carefully considered as traditional buildings often have many regulatory stipulations and have to perform in circumstances in which the breathability of the overall building element is critical when trying to prevent deterioration due to dampness or ingress of moisture. Effesus, a European research project, is looking to use insulating products such as aerogel to dramatically improve the insulation properties of plasters (2013). This project predicts that 'Europe can become the leader in CO_2 emissions reductions by applying innovative solutions to its built cultural heritage' (Effesus 2015).

Whilst revolutionary aerogel-based super insulants are appealing, the ability to renovate and repair existing parts of buildings with minimal intrusion at the same time as improving the thermal and durability performance of the property is an obvious, though less glamourous, next step. The key to products that will allow for this is to offer other advantages to the designer/specifier beyond the technical performance, such as minimising internal disruption within a renovated building and reducing the need to decant occupants. Simple but effective innovation in these circumstances can be provided by the use of over-roof insulation for profiled roofs. These systems can also offer an opportunity to extend the life of roofs that are considered beyond economic repair and thus effectively extend the life of the whole building. As usual, with any insulation, the interaction of the moisture properties of the upgraded fabric needs to be balanced with the improved thermal performance.

Other innovations based on non-food crops, such as straw-bale buildings, have been mentioned in Chapter 4, but future uses are likely to look to a combination of off-site construction and natural materials. An example of this type of innovation is straw panel technology, such as that produced by ModCell (in collaboration with the University of Bath) and Ecofab (2014). The ModCell system was used in a new housing development in Leeds and monitored in relation to its energy use. The residents living with ModCell have seen their energy bills reduced by up to 90% compared to the city average after spending six months in their new homes made from straw bale panels (Timber Expo 2013). Both ModCell and Ecofab construct their panels in a controlled environment in a modular format, increasing the quality of the product compared to site-based construction methods. Placing the panels on site is very straightforward, adding to the ease of use for future sustainable construction projects.

Renewable technologies are an obvious area in which new advances can be made. The efficiency of some sustainable technologies is currently quite poor, and this problem applies to photovoltaics (PVs) in particular. New PV technologies include a panel that is 20% transparent to all light wavelengths and can be used as normal glazing at the same time as it generates electrical power. Polysolar is one company working with the latest organic polymer PV materials, technologies and processes to manufacture PV windows (Polysolar 2013).

As more futuristic developments in materials gather pace, unique substances such as graphene could be one of the materials that, through its properties, changes the applicability of many existing products used in buildings. Graphene is a single layer of carbon atoms arranged in a hexagonal lattice and it has unique mechanical, electrical, and optical properties. Various international research organisations are investigating possible new products made with graphene or improvements on existing products. One such innovation is based on combining graphene with another atom-thick material to create paper-thin solar surfaces. This process could create a graphene solar coating for powering buildings when applied to exterior walls (Daily Mail 2013). A series of projects in universities based in the UK is in the process of researching how to enhance the 'manufacturability' of graphene:

- Scientists from the University of Cambridge are researching flexible graphene electronics and optoelectronics, which could include touch screens.
- Imperial College London is investigating the potential of engineering with graphene for multifunctional coatings and fibre composites, and graphene three-dimensional networks.
- Manchester University and the University of Cambridge are jointly exploring the use of graphene for super-capacitors and batteries in energy storage applications. These might have direct applications in dealing with the tricky problem of peaks in electrical generation from renewable technologies in buildings.
- Proctor and Gamble and Dyson are collaborating with Durham University to explore applications of graphene composites with a view to producing thinner, stronger material applications.

These projects will continue to progress over the coming years and, alongside work across the globe, new products will probably be influencing the construction industry in five to eight years' time.

9.5   Future construction practices

The way in which the construction industry works is also very likely to change in a great many ways. This section of Chapter 9 looks at likely future influences upon construction professionals, bearing in mind the probable directions of process-related developments.

Building information modelling (BIM) is currently one of the driving forces of reform in terms of process and practices related to construction (see Chapter 6 for more information). The exact characterisation of how BIM will look or how long it will take to make headway in the construction industry is not certain. However, as described in the previous chapters, with governmental backing, appropriate standards that regulate its use (alongside a clear view of the benefits of BIM) mean that the use of BIM will increase. Therefore, a chapter that professes to look at the future of sustainable construction and does not take into account the future practice-related impacts of BIM would be unsatisfactory. One aspect of BIM that is likely to grow is the integration of new streams of information. Currently, environmental impacts related to construction activities and buildings in particular are very much stand-alone operations, but the future extensive use of BIM could integrate and widen the scope of this area as it relates to the responsibilities of construction professionals. Data related to the environment can be BIM integrated using measures such as probable life of components, reminders from suppliers that new more efficient versions of products are available, advice for building management staff throughout the lifetime of the building, and finally end of life advice for reuse, recycling, and disposal. The UK's BRE has examined this aspect of future use of BIM through a project entitled IMPACT. This is a specification and dataset for the incorporation of building-level embodied impact assessment and life cycle costing into BIM applications. More information is available on the impact website: www.IMPACTwba.com. Further innovations are likely to include taking BIM compatibility beyond all computer-aided design (CAD) systems (extending existing developments in that field) to simulate intelligent energy solutions. This might take the form of an evolved version of immersive systems such as Sefaira, which provide communication data sets alongside comprehensive simulations of holistic building performance. Graphic BIM communications are likely to be the main method of interaction, allowing fast and efficient grouping of decisions, integrating performance analysis. Planned off-site refurbishment is also likely to benefit from being BIM compliant and from integrated systems. The variable dimensions, materials, and design aspects of existing buildings provide a difficult problem when trying to match these buildings with modular and more uniform panels. However, if BIM-related 3D models can be linked to appropriate manufacturing processes and facilities, then customised upgrade panels can be efficiently produced and placed.

An instance of the use of high-performance off-site refurbishment has been demonstrated at the Parkview Hub residential refurbishment development on the Thamesmead Estate, London. The residential units are five storeys high and being refurbished by Gallions Housing Association to the Passivhaus Institute–certified EnerPHit standard. The retrofit used a prefabricated timber cassette approach, developed by contractor Gumpp & Maier UK, which is assembled in a Bavarian factory. The refurbishment was completed at the end of 2014, and it is likely that this scheme will act as an exemplar for energy-efficient refurbishment, influencing future similar projects (Retrofit Roadshow 2013).

Offsite production is one method of introducing forward thinking in relation to construction processes; however, the need for land might throw up different problems. Much of Hong Kong Harbour is built using reclaimed land, as are many parts of other prominent cities. A competition to develop a design for a floating village of six hectares for the Royal Docks has been put forward by London's mayor. Floating developments do exist elsewhere, for example in Ljbury near Amsterdam. The London proposal would have an impact in terms of reviving the commercial aspects of the waterways and could have other communal benefits (Monaghan 2013).

Leased properties are, to some extent, under the radar. Marks and Spencer (M&S) is one organisation that uses this kind of agreement to promote sustainable development. It is introducing green clauses into leases for both new and existing stores. As part of the M&S Plan A programme, green clauses will facilitate sharing of waste information and data such as gas, electricity, and water usage in M&S-occupied buildings to encourage the retailer and its landlords to make greater carbon reductions (Marks & Spencer 2013). It also encourages a joint approach to investment in eco-building technology, such as biomass boilers, LED lighting, and rainwater harvesting.

Future partnering practices related to the decisions that lead contractors to source their labour as well as their materials locally are likely to change. This social aspect of sustainable construction can influence clients in terms of tendering or partnering processes. An example of this is apparent in the agreement at the Birmingham City University campus in the UK. Willmott Dixon (the main contractor) has committed to spending over half of the contract sum with local subcontractors and suppliers, focusing its local spend on firms within a 20-mile radius of the project (Link2 2013). This fits with the client's defined goals that include spending considerable sums locally.

9.6   Future norms and expectations

Future building occupants will have expectations related to how their buildings will perform. In an age when our phones, televisions, and even our washing machines are 'smart', shouldn't our buildings be more interactive? The reality is that a balance has to be struck between advances in the use of embedded technologies and the ability of the average occupant to make the most of these technologies. If we make things simple and intuitive to use, the elderly and those of us who are less comfortable with technology will still find ourselves able to receive the same service from our buildings as those who are more technologically able. Currently, as can be seen from the findings in Chapter 7, building occupants, particularly householders, sometimes have difficulties getting their homes to work efficiently for them. A small study focused on the relationship between occupants and the technologies incorporated into six sustainable homes (NHBC 2012). In this study the NHBC Foundation found that the occupants were not persuaded to purchase their property because of the low-carbon technologies that were incorporated into them. The occupants stated that they rated the householder guidance supplied with such technologies as 'unsuitable'.

The recommendations from the report critiqued the operation of a sales team for low-carbon homes:

- The sales staff had a very limited ability or willingness to communicate the benefits and opportunities of low- and zero-carbon technologies in a way that the prospective purchaser could understand or be inspired by.
- There was a varied level of understanding among sales staff about the specific low- and zero-carbon technologies being fitted in the homes.
- Homeowners were not routinely invited to provide feedback to the sales teams post-occupancy on their experience of the use of low- and zero-carbon technologies.

To equip sales staff appropriately so that they can advise prospective purchasers developers should instil in them that sustainable technologies need special care and attention in comparison to those in a conventional property. Within reason, any future occupant should be able to move into a property in which the installed systems 'work'.

This conflict between the promise of what technology, practices, behaviours, and policies can provide and what happens when they are used by everyday occupants rather than in exemplars, pilot studies, or best practice projects is the crux of the relationship between sustainable construction and sustainable living. As this text proves, the technologies have been researched, built, and tested, and they work. The practices have been examined and recommendations made. Policies have been set down, but slippage is occurring. The major reason for the lack of real progress can be related to the difficulties of scaling up these changes so that, in a busy world, everyone who can live in sustainable buildings will be able to.

9.7   Chapter summary

The aim of this book has been to provide some guidelines for the sustainable construction of buildings in the current climate: literally and metaphorically. Definitions of sustainability—and the extent to which we can confidently assert a building has been designed, constructed, operated, and occupied sustainably—will continue to be debated and contested. Yet a lack of a prevailing consensus or the persistence of an uncertain threshold for success should not release us from the task of attempting to find new and more sustainable ways of constructing and managing our built environment. As we have seen, the multiplicity of disciplines that can contribute to sustainable practice include mathematics, politics, management, science, accounting, art, design, construction, law, and psychology. These—and ever new advances in each—continue to shape our buildings in unforeseen ways with unknown potential for change in the future. Indeed, given the rate of change in the construction industry and wider advances in knowledge and technology, it is possible, even likely, that within a decade sustainable construction will become a mainstream rather than niche interest within the construction industry. Until that time it is hoped that this volume will contribute in a small way to that shift in thinking and practice towards a more sustainable future for construction.

References

BGS (2013) *Heat energy beneath Glasgow*, Research at British Geological Survey (BGS), [Online], Available: http://www.bgs.ac.uk/research/energy/geothermal/heatEnergyGlasgow.html [23 May 2013].

BP (2013) *BP Statistical Review of World Energy June 2013* [Online], Available: http://www.bp.com/content/dam/bp-country/fr_fr/Documents/Rapportsetpublications/statistical_review_of_world_energy_2013.pdf [9 Sept 2014].

BSRIA (2013) News, December 2013. [Online], Available: https://www.bsria.co.uk/news/article/connected-and-smart-home-market-will-reach-620-million-in-europe-by-2015/?utm_source=http%3a%2f%2fbsria-marketing.co.uk%2fbsrialz%2f&utm_medium=email&utm_campaign=BSRIA+E-News+Dec+2013&utm_term=Part+L+changes+l+IAQ+issues+l+Co-heating+test+reliability&utm_content=20219 [23 May 2014].

Buro Happold (2013) *London's Zero Carbon Energy Resource: Secondary Heat Report Phase 2*, City Hall, Greater London Authority, London. [Online], Available: http://www.london.gov.uk/sites/default/files/031250%20(final)%20GLA%20Low%20Carbon%20Heat%20Study%20Report%20Phase%202.pdf [19 Dec 2013].

Buscher, V., Doody, L., Tomordy, M., Ashley, G., Tabet, M., & McDermott, J. (2010) *Smart cities: Transforming the 21st century city via the creative use of technology*, Arup. [Online], Available: http://publications.arup.com/Publications/S/Smart_Cities.aspx [7 Oct 2015].

Cheshire, D., & Grant, Z. (2007) *Sustainability, CIBSE Guide L*, CIBSE, London.

Daily Mail (2013) Graphene breakthrough could mean buildings powered by solar paint. Published 15:10, 26 April 2012 [Online], Available: http://www.dailymail.co.uk/sciencetech/article-2135571/The-end-bulky-solar-panels-New-solar-paint-generate-electricity-roofs-walls-EVERY-home.html [11 April 2014].

DCLG (2012) *Bringing Empty Homes back into Use*, DCLG, Department for Communities and Local Government. [Online], Available: https://www.gov.uk/government/uploads/system/uploads/attachment_data/file/5926/2073102.pdf [23 May 2014].

Ecofab (2014) About us—Technical. [Online], Available: http://www.eco-fab.co.uk/pages/technical [23 May 2014].

Effesus (2013) Researching energy efficiency for European historic urban districts. [Online], Available: http://www.effesus.eu/wp-content/uploads/2013/07/Effesus-Leaflet_English-reduced.pdf [19 Dec 2013].

Effesus (2015) Researching energy efficiency for European historic urban districts. (Quote from Home page line 1), [Online], Available: http://www.effesus.eu/about-effesus

EHN (2013) *Empty homes nationwide*. [Online], available: http://ehnuk.org/ [19 Dec 2013].

Gething, W., & Puckett, K. (2013) *Design for Climate Change*, RIBA Publishing.

Harnmeijer, J., Parsons, M., & Julian, C. (2013) *The community renewables economy: Starting up, scaling up and spinning out*, ResPublica Trust. [Online], Available: http://www.respublica.org.uk/documents/yqq_Community%20Renewables%20Economy.pdf [19 May 2014].

IBA (2013) Internationale Bauausstellung IBA Hamburg (International Building Exhibition) [Online], Available: http://www.iba-hamburg.de/en/iba-in-english.html [19 Nov 2013].

Innovate UK (2015). [Online], Available: https://connect.innovateuk.org/web/design-for-future-climate [7 Oct 2015].

ISO (2014a) First draft of ISO's standard for sustainable communities reaches CD stage [Online], Available: http://www.iso.org/iso/home/news_index/news_archive/news.htm?refid=Ref1877 [20 Nov 2014].

ISO (2014b) *ISO/CD 37101 Sustainable development and resilience of communities—Management systems—General principles and requirements*. [Online], Available: http://www.iso.org/iso/home/store/catalogue_tc/catalogue_detail.htm?csnumber=61885 [20 Nov 2014].

ISO (2014c) *ISO 14046:2014 Environmental management—Water footprint—Principles, requirements and guidelines*. [Online], Available: http://www.iso.org/iso/catalogue_detail?csnumber=43263 [20 Nov 2014].

Link2 (2013) Willmott Dixon develops creative streak. [Online], Available: http://www.link2portal.com/willmott-dixon-develops-creative-streak [23 May 2014].

London Mayor (2013) Waste heat from the Tube will help to warm hundreds of homes. [Online], Available: http://www.london.gov.uk/media/mayor-press-releases/2013/11/waste-heat-from-the-tube-will-help-to-warm-hundreds-of-homes [19 Nov 2013].

Marks & Spencer (2013) Plan A. [Online], Available: http://plana.marksandspencer.com/about [19 Dec 2013].

Met Office (2011) Climate: Observations, projections and impacts, UK Met Office Hadley Centre DECC. [Online], Available: http://www.metoffice.gov.uk/media/pdf/t/r/UK.pdf [23 May 2014].

Monaghan, A. (2013) Boris Johnson reveals plans for UK's largest floating village, *The Telegraph*, 12 Mar 2013. [Online], Available: http://www.telegraph.co.uk/finance/newsbysector/industry/9925924/Boris-Johnson-reveals-plans-for-UKs-largest-floating-village.html [11 Dec 2013].

NASA (2005) Climate and global change website, updated 02.01.05. [Online], Available: http://www.nasa.gov/mission_pages/noaa-n/climate/climate_weather.html [12 May 2014].

NHBC (2013) *Foundations report NF 50, Designing homes for the 21st century, Lessons for low energy design*. [Online], Available: http://www.nhbcfoundation.org/Researchpublications/NF50Designinghomesforthe21stCentury.aspx [23 May 2014].

NHBC (2012) Low and zero carbon technologies, (NF53) NHBC Foundation. [Online], Available: http://www.nhbcfoundation.org/Researchpublications/LZCtechnologies.aspx [12 May 2014].

Poincaré, H. (1913) *A Series of Volubles for the Promotion of Scientific Research and Educational Progress*, Vol. **1**, (Ed. McEeen Cattell, J.), The Foundations of Science: Hypothesis in Physics, Stanford University libraries, The Science Press, New York, p. 129.

Polysolar (2013) Future innovations. [Online], Available: http://www.polysolar.co.uk/future-innovations [19 Dec 2014].

RAE (2013) Smart buildings people and performance, Royal Academy of Engineering, London. [Online], Available: http://www.raeng.org.uk/news/publications/list/reports/RAEng_Smart_Buildings.pdf [19 May 2014].

Retrofit Roadshow (2013) News. [Online], Available: http://www.retrofit-roadshow.co.uk/news/industry-news/offsite-construction-key-to-major-uk-passivhaus-refit/ [23 May 2014].

RICS (2013) *Modus*, The Oil Issue, UK Royal Institute of Chartered Surveyors (RICS) magazine, 31 October 2013. [Online], Available: http://www.rics.org/us/knowledge/journals/modus/recent-editions/the-oil-issue/ [24 May 2014].

Sainsbury (2014) Sainsbury's 20x20 Factsheet: 01 Respect for the Environment. [Online], Available: http://www.j-sainsbury.co.uk/responsibility/20x20/ uploaded 25 March 2014 [13 May 2014].

Satterthwaite, D. (2007) The transition to a predominantly urban world and its underpinnings. International Institute for Environment and Development. *Human Settlements Discussion Paper Series*, 4 September 2007.

SHARC (2014) A hot resource. [Online], Available: http://www.sharcenergy.com/ [20 Nov 2014].

TEDGlobal (2011) Wireless data from every light bulb. Ted Talk from Ted Haas. [Online], Available: http://www.ted.com/talks/harald_haas_wireless_data_from_every_light_bulb.html [19 May 2014].

Timber Expo (2013) News. [Online], Available: http://www.timber-expo.co.uk/media/industry-news/straw-bale-homes-save90-on-energy-bills/#.U4NzX9JdXT8 [19 Dec 2013].

UK Government (2014) Policy. Increasing the use of low-carbon technologies. Supporting detail: Heat networks. [Online], Available: https://www.gov.uk/government/policies/increasing-the-use-of-low-carbon-technologies/supporting-pages/heat-networks Updated May 2014 [19 May 2014].

UN (2012) *World Urbanization Prospects, The 2011 Revision, Highlights,* ESA/P/WP/224, Department of Economic and Social Affairs, Population Division, United Nations, New York.

United House (2013) News. [Online], Available: http://www.unitedhouse.net/about-us/news/id/1382369956 [23 May 2014].

Veolia (2013) Imagine 2050: The innovations we will need in waste, water and energy to ensure a sustainable future. Creative solutions for our environment, Veolia Environment. [Online], Available: http://www.veoliaenvironmentalservices.co.uk/Documents/Publications/Main/Imagine%202050/Imagine%202050%20Brochure.pdf [23 May 2014].

Webb, M. (2013) Brilliantly bizarre ways to power our lives, MSN Innovation. [Online], Available: http://innovation.uk.msn.com/future-of-transport/brilliantly-bizarre-ways-to-power-our-lives uploaded 23/12/2013 [24 May 2014].

Wolf Jr., C. Bamezai, A., Yeh, K., & Zycher, B. (2000) Asian Economic Trends and Their Security, Rand, Santa Monica, California (Chapter 1, quote p. 1).

WWT (2014) Wessex Water's wastewater metering trial is a world first, *Water and Wastewater Treatment.* 29/07/2014. [Online], Available: http://wwtonline.edie.net/news/wessex-water-s-wastewater-metering-trial-is-a-world-first#.VG230CLkfYh [20 Nov 2014].

Index

Sustainable Construction Processes: A Resource Text, First Edition. Steve Goodhew.
© 2016 John Wiley & Sons, Ltd. Published 2016 by John Wiley & Sons, Ltd.